International Texts in Critical Media Aesthetics

Vol. 9

Founding Editor:
Francisco J. Ricardo

Series Editors:
Francisco J. Ricardo and Jörgen Schäfer

Editorial Board:
John Cayley, George Fifield, Rita Raley,
Tony Richards, Teri Rueb

INTERNATIONAL TEXTS IN CRITICAL MEDIA AESTHETICS

The Internet Unconscious

On the Subject of Electronic Literature

SANDY BALDWIN

Bloomsbury Academic
An imprint of Bloomsbury Publishing Inc

B L O O M S B U R Y
NEW YORK · LONDON · OXFORD · NEW DELHI · SYDNEY

Bloomsbury Academic

An imprint of Bloomsbury Publishing Inc

1385 Broadway	50 Bedford Square
New York	London
NY 10018	WC1B 3DP
USA	UK

www.bloomsbury.com

**BLOOMSBURY and the Diana logo are trademarks of Bloomsbury
Publishing Plc**

First published 2015
Paperback edition first published 2016

Library of Congress Cataloging-in-Publication Data

Baldwin, Sandy, 1966-The Internet unconscious : on the subject of electronic literature /
Sandy Baldwin.pages cm.– (International texts in critical media aesthetics ; vol. 9)
Includes bibliographical references and index.ISBN 978-1-62892-338-4 (hardback)
– ISBN 978-1-62892-340-7 (ePub)– ISBN 978-1-62892-339-1 (ePDF)
1. Online authorship. 2. Hypertext literature--History and criticism. 3. Literature
and the Internet. I. Title. PN171.O55B35 2015302.23'1–dc23
2014035873

ISBN: HB: 978-1-6289-2338-4
PB: 978-1-5013-2001-9
ePub: 978-1-6289-2340-7
ePDF: 978-1-6289-2339-1

Series: International Texts in Critical Media Aesthetics

Typeset by Integra Software Services Pvt. Ltd

CONTENTS

FOREWORD

As an opening comment, may I first say that there's really one book whose optic is similar to this one, combining the received understanding of armatures—physical supports and their precedents—with the potential of their function going in a different direction than through a discussion of their effects. That distinct direction traces, instead, an exploration of their existence as *metaphors* of their function, and these metaphors are really intuitive forms of embrace that I'll clarify later.

This book works through the operation of a mechanism back to a purpose and then takes the offramp of subjective reflection of both. In this new direction, the exploration of a system acknowledges but is not defined by its legacy. The personal dimension takes expressive primacy, and therefore, the text can work through the details of a medium's function and still retain the ineluctable whisper of personal meaning, of which so little has been said in the history of literary or even visual criticism. If we wish a quick label, then we can say of Baldwin's writing that it weaves an inviting prose of poetic phenomenology, and, as I say, I have seen it fully developed only in one other text—Rosalind Krauss's *The Optical Unconscious*. As Krauss challenged the story of modernism in art, with its assumptions about a steady development, a consensual view of what matters in the evolution of that very special medium-specific reflection that accompanied all modernist movements, Baldwin, too, takes on the assumptions—which is to say, the preference of function over potential—in the project of the net, or the net as a project. By unveiling some of the texture of its meta-expressive strata, the net is made manifest as the most sweeping work of modernism possible. Here, for instance, is modernism in the flesh, considering its condition:

Everything is an instrument towards the end of the spectacle. I am a project organized by spectacular desire. It is a spectacle for me. To gaze on the other is to consume and incorporate. I hear a movement, the floor creaking behind me, and suddenly I am caught—in this moment the entire situation is reversed. I am an object of scrutiny. If in the first case, everything was given over to the project of taking the other in, in the reversal I am displayed and pinned in position for the other to consume. The other is localized, captured, and introjected. I am put on display and grovel under the other's gaze. (89)

As he brings us to it, Baldwin's aim makes the mark move, the text of the net is a performative verb, and he brings us to those marks, in this new Lascaux, to see the "writhing on the wall" (107), that is, advancing it as a literary case, without putting it literally. Given that the net has been identified as the opposite—something literary rather than literal—the book, righting our distortion, is a tour de force. Let us recall this précis of the work's line of attention: "Net writing is no longer 'elsewhere' but rather brings its elsewhere and otherness to everywhere."

So much for the manifest elements of Baldwin's focus: the net, writing, and the mechanism between both. More fundamentally, the impact of this book should be understood in light of two modernist global experiments that have colluded and consolidated into a unitary kind of wavelength: that of the maturity of literature and that of the adolescence of the digital network. Each can be seen as an instrumental phenomenon in self-expression. But what has been less clearly explored is the question of the *subject* within the scope of expressive output. As it happens, what is identified by the term "subject" is now not merely a creative individual, but an instrumental institution or layer whose semiautonomous agenda has invaded, infused, and redefined preexisting assumptions of expressive agency. The separating line between speech and writing that Walter Ong's *Orality and Literacy*, Marshall McLuhan's *The Gutenberg Galaxy*, and others traced follows, through a distinct tributary established through Derrida's *Speech and Phenomena*, and earlier still, Husserl, and even Rousseau. Written language transformed language (which is to say, orality) and legitimized the membrane that gradually altered history from inclusive, participatory lore into a class commodity exchanged by one ideological elite after another.

Considering the impact of this expressive bifurcation, the paucity of reflection on its emergence may seem striking, but it is not: the meta-reflective act of commenting on the expressive apparatus itself, rather than experiencing what comes from within its frame, as the essence of modernism, escaped systematic study until the nineteenth century. Less obvious still is the fact that expressive forms were themselves undergoing profound transfiguration, so that any commentary on their structure became itself, from Mallarmé to the Dadaists and thereafter, promoted to a category of expressive content. It would seem pointless to rehearse the many exploratory attempts toward structure-as-content of the literary enterprise, largely because they were developmentally incomplete, but also because they understood "literature" to be a phenomenon that was print based, as it was until a few decades ago and still remains today.

The larger concern, however, is that while studies of literary form approached their study through attempts at some formal method (Tzvetan Todorov and Northrop Frye come immediately to mind), they needed to freeze the literary act into a subject-less, static practice amenable to comparative appraisal through the incessantly reductive monstrosity known as genre. In turn, the acceptance of genre as a central axis point in literary expression subverts and undermines deep discussion of the subject because structural-formalist thinking is, as with all structuralist deficits, required to reduce and equalize the truth of *creator*, *created*, *creation*, and *creative witness* to approximate variables, when we understand, by the simplest reflection on these terms, how distinct their existential universes really are, and how it is that, by this difference, all literary expression derives its power. This is also why structural and formalist analysis cannot be spoken; it can only be written—it is part of the elite tradition that erases the participatory realism of the subject in any adequate measure, experience being exchanged for expediency.

To grasp the literary, particularly in a new non-static form, as it flourishes within the digital network, both literature and its structure must be re-symbolized, re-infused with what comes to the formation of imagination, impression, and experience; the boundary assumptions of the subject in the net, and the literary in general, must, as understood until now, be questioned and ultimately surpassed; the subject in the machine, in the story, in the imagination of an authorial protocol can only be understood when its being communes with that of the reader—itself a deeply problematic and condescending term. An entire panoply has arisen to foment the mendacious distortion that separate entities are somehow in play in the net and in all writing, even in speaking. There cannot be a subject in the literary work and another subject reading it, and yet a third subject trying to arbitrate among both through some analytic scaffold, and a fourth subject reading the analytic subject's attempts at distinguishing, conjoining, or even refuting the other two subjects. Baldwin reimagines the underlying communion between all these false assumptions of expressive separateness—including that of the system that produces the expressive signs itself—by a simple, crucial, and necessary move: he obliquely, but persistently brings us into his world, which is actually ours, and verifies with us (because we are one in all of these roles) that "subject" is not a plural term, and that any distinction or plurification of the term is misleading. If there be experience in any, then all are in our being and vice versa. That is the net behind every collective consensus of individually expressive organization, regardless of its existence in machinery or in imagination—it symbolizes our inability to separate meaning, and as such it is incessantly read and written.

Let us now leave it at that, and let his writing unravel the partitions that we have naively accepted between the idea of distinct worlds between the fictive, the linguistic, the readerly, literary, and the machinic. All of these are networks of signs encased in a larger network of meaning, none of which exists in any particular way without our own act of presence that forms the inseparable horizon of imagination and experience and whose electronic version is a mechanism, a medium, and a dreamcatcher.

Francisco J. Ricardo
April 2014

INTRODUCTION

There is electronic literature that consists of works, and with this the authors and communities and practices around such works. This is not a book about that electronic literature. It is not a book that charts histories or genres of this emerging field, not a book setting out methods of reading and understanding. Existing criticism posits electronic literature as works that are different from but part of the apparatus of networked digital writings. In turn, the writers of such works are subjects operating with the tools of the networked computer in ways that are different from but part of this apparatus. The apparatus of networked digital writing posits, locates, and regulates these works and subjects.[1]

The apparatus operates through writing in the form of codes and transformations of codes. Everything on the computer is writing. Everything on the net is writing in sites, files, and protocols. Everything is well-formed and complete. The structure of writing on the network is the structure of all things, all ported and downloaded and uploaded: the triumph of informatics. Literature, as literally "what is written," is an exemplary case of such writing within the apparatus. It handles the expressive and creative potential of the subject within the larger managerial framework of networked writing. In this framework, the value of a discourse on electronic literature is its elaboration, containment, and limitation of this potential. Electronic literature is a dutiful part of the bureaucracy, and all existing discussions of electronic literature imply some version of this managerial logic.

This book takes a very different approach and seeks out a different electronic literature. I start from the basic claim for an imaginary overinvestment in writing on the network that makes every inscription a flickering presence and absence of my body and the other's body. Such writing involves hysterical and ecstatic repetition, abjection, and impossibility.[2] I focus on electronic literature as both the name for this writing and as its residue or detritus. Rather than the belletristic corpus of electronic literature, this approach deals with the problematic "literary" at work in all networked writing.

I differentiate between two ways the literary appears and is read in networked writing. The first involves hysterical and ecstatic repetition, abjection, and impossibility. I call this "love of literature." Love of literature is the delirium of bodily transcendence where the screen becomes "you." Electronic literature from this perspective is a discourse on this imaginary and a reading of its work. We search for transitivity and translation. We seek for something that comes across in the screen. We desire the screen to be the face. In electronic literature we read this survival, but only as a fiction, only as literature.

The second involves "the politics of the literal," or the literalization of the imaginary, and is the content and subject of what I termed managerial logic above. Imaginary overinvestment drives managerial logic, control mechanisms, and the well-formed objects we encounter on the net. The network is maintained by our investment in reading and writing it. To read writing on the networked computer is to repeatedly stage an encounter with otherness and interiority that never quite takes place. To discuss the literary on the network is to read the work of this repetition. The works of literature are the detritus of the other's body that cannot be reached, can only be read and re-read. Electronic literature from this perspective is a symptom of an impossible situation.

In the presence of another person, I face a demand of recognition: the Levinasian face of the other demands I respond. The other demands that I stay, that I witness, that I love and die with the other. The other may face me and demand my submission, or may demand conflict as an enemy. A screen of digital writing is very different: it is a fuzzy and displaced demand. Neither presence nor absence of the other, digital writing involves a distance that I repeatedly cross by reading and writing. Politics occurs at the screen: bodies are situated and held in place; subjects are required to sign, to click, to look, and to answer. The politics of the screen are where an immobilized body is stopped and where precisely this stoppage releases the imaginary into the network.[3]

To discuss the networked computer in terms of sites, files, protocols, and so on, posits a domain of logically ordered, fully formed objects. It posits writers and readers that survey and take in this domain. Such a domain is not possible and our gaze on the screen of our networked computers is pushed to the edge of philosophical logic and certainty, where subjectivity is abjected and where objects are ruptured. The net is impossible and intractable. The Internet "stack" leaps from hardware to application to display. The Internet Protocol (IP) domain of net packets remains a pack without direction and place, only committed to increase and flow. The relation is unstable, a "leaky abstraction" in the words of Joel Spolsky.[4] In such a field of messiness and flux, the only possible discourse on the networked computer is one of impossibility. Everything on it must be considered as if one long invention. The Internet is a work of literature.

My chapters read this narrative and follow its poetics, follow what is invented in networked digital writing. I describe a series of site and encounters, everyday and familiar to all: email, logging in, the screen itself, and so on. In every case, I show how the encounter with digital writing and its demand leads to the double play of literalization and imaginary overinvestment. The literary of networked writing—its aesthetic interest, its work—is situated in this double play. The chapters that follow move from writing as a solitary explosion of hope and desperation to works of writing the net that unify others as friends, lovers, and communities. Earlier versions of some of the material in this book appeared in *symplokē*, *Formules*, *Performances Research Journal*, and *Intimacy Across Visceral and Digital Performance* (Palgrave Macmillan, 2012).

Many inspirations led to this book. I also thank Jörgen Schäfer, John Cayley, all the editors at Bloomsbury, and all those others who supported my work. I especially thank Francisco Ricardo, founding editor of this book series. Among many things, he wisely pointed me to the appropriate title. The link to Krauss' *The Optical Unconscious* is so clear that it remained a blind spot for me until Francisco's intervention. Re-reading Krauss after many years led me to recognize crucial aspects of my project and also pointed to where I might continue beyond this book. I was already thinking of Alan Sondheim's "ASCII unconscious," which he describes as "part extension, part catalyst, part resonant response to the other, only part of which is consciously deliberated; it forms a partial mapping of desire and partial objects; it replaces the real, insofar as communication is online."[5] The partialities in this description are very much at work in my writing: the dreamed hope of replacement objects, mappings, and communications is what I find in electronic literature. Alan inspires my thinking through conversations and collaborations for many years now.[6] The work of another philosopher, Alphonso Lingis, showed me the possible imperatives played out in the things of this world and the way they are said and lost in writing.[7] I feel lucky to have emailed with Lingis and exchanged some thoughts.

These thinkers, readers, and friends all inspire what follows, but I dedicate the book to Kathleen.

I

As if I wrote the Internet

As if I wrote the Internet, on my iPhone, wrote the entire thing, texted it, 140 characters at a time. As if I wrote it. It, what is it? The word suggests the thing that language does not touch. It is the totality of the net, that impossible total text.

As if I wrote it yesterday, in my spare time. Why not? The net seems eternal, timeless, but also renewed and endlessly young, constantly written and re-written. Why not? The Internet seems personal, so mine, ranging from the pages with my name on them—and they are everywhere, I find my name in every page, I read it between the lines, on CNN dot com, on blank pages, on every page everywhere—from such explicit namings, to the oddest, most extreme corner of the net. Of course, there are no corners; the net is shapeless and all shapes: it touches itself and touches me. It is part of my pathology. I am written and rewritten into and by it. Every page, every code, every character, every email, every blog post, every YouTube video, every torrented song. I wrote it all. I am modest, but I must take credit. That Facebook status update and Tweet you just sent? I wrote those as well.

But you are a writer, too. Perhaps you wrote the Internet? It could be, I give you that, it could be otherwise, it could be written by you, as if.

"I write this" or "I wrote this." The letters appear on screens, perhaps on your screen. How could I do this? Is it not almost a crime, a crime to write that "I wrote this," to claim the writing on the screen, in that great beyond. Is it not almost a crime against humanity? To write, the inscriptive act is the end of humanity and the beginning of something other. Who am I, who are you that this can take place? What took place?

The "oddness" I feel is the unbridgeable gap, as I face these letters on the screen. Oddness or delirium that these words are mine. Why mine? How could any writing belong to me, or to anyone? A crime to even imagine—the very thought!—of leaving a trace that is mine, of the impossibility of these smooth and abstract characters as somehow by me. Truly, this is a delirium or hysteria of nothing. An insanity of writing. It must be insane to call these words mine.

Whatever else it is, literature is the possibility of uncontrolled enunciation, of inscription without a trace, of inscription in and as all traces, everywhere as if inscribed. "Inscribed" is a term too easily deployed: not simply marked but marking as an act, marking me in the mark. Inscribed is both "it is written" and "it writes," both "I am written" and "I write." Inscription is an ambient philosophical given—"there is" inscription. The literary critical readings that thematize inscription simply describe aspects of this ambient condition.[1] Inscription is also tied to a narrow range of very specific scenes and sitings, where otherness is announced through inscription, sitings, and scenes such as those described later in this book: the proper name, the avatar, the address, permission, and so on. The slightest or the largest act is inscribed: a click of a mouse or the entirety of the net. The first is almost nothing at all; the second a totality I cannot imagine, one that saturates all imagining.

I confess not to care about this or that work of electronic literature. It is both a blindness and a cultivated response. I cannot become interested and I refuse to. I work hard at this indifference, seeking a general recognition of the literary as it takes place in digital writing, indifference to individual works and openness to the work of writing.

It is the subject of electronic literature that concerns me, a subject at work and worked on in scenes of reading and writing. I try to resist the determination that some works are more interesting or some less—some works on one side, some on the other. Literature befalls writing. It takes place and takes the place of the mark; it takes it to another place. The other place or the great beyond is where the concern with the subject of electronic literature becomes a concern for others. In the absence of the trace, in electronic writing utterly bereft of my presence, there is literature. Electronic literature is a consequence of this fact, and my project—here and everywhere—is to discover this consequence. Consider this—my book, my presence before you—as a performance of the problem of literature. There are paths and a problem to follow toward electronic literature. Indeed—some of these paths and problems are named, are works of e-lit. I cannot say the totality of the net; I can only say many things as I follow the problem of literature. My mission: offer a tribute to and recognition of the problem.

It is the institution of the name and the text, or better of the login and the file, that regulates our inscriptions, that indebts us to writing online, such that I assert possession of certain web pages and you other pages, and together we form a crowd of electronic writers. Literature is the seal and guarantee of such inscription. It is the consumption and collapse of this seal. I wrote every work on every computer everywhere and for all time. I put my name to it all, just now. Literature's "everything is permitted" means acceleration, not regulation.

The great beyond

Imagine you are writing on the computer. Where is this "on"? On the screen, but not "on" the screen. The sense certainty that enchained the graphite of the pencil left on the page to your writing hand or chained the impact of the typewriter key to ink on paper is not the case here. You are not writing *on* the screen though letters may appear *there*. Even if you touch the screen, as I do with a smart phone, your touch hovers above the screen; my touch touches the glass that separates me and my world from the screen. There is the screen that I touch: it is glass, it is chrome, but the words are deeper; they are not on the screen. Fanning my fingers on the screen, touching its cool glass, I get nothing but smooth surface. Licking the screen, rubbing my chest on it, none of these acts brings me closer to the other side. Do not talk to me about haptics or VR or what have you, all of which only reinforce the gap. The membrane is absolute. They are on the other side of the screen in *the great beyond*, in that *other place* seen through the screen, where my letters and words and messages accumulate. We all know this experience: to speak of a web page or email or any form of telecommunication is to acknowledge this beyond, this deepening of the screen that is never "here"—there is no depth to measure, no jumping off point—but is always "there." Because of this, I ask that you imagine writing on the computer; "imagine" because digital writing involves such an imagining, involves such imaging.

I underline and enforce this great beyond because it is too easy to again return to discuss electronic literature and digital writing, to fall back into the easy discussion of such writing and literature as here and given instead of there and impossibly distant. Instead, I stay fixed and entranced by the oddness of this break.

I write on the computer, putting words in this other place. The screen shows and displays mistyping, errors. The screen is an image that I read. Whatever else it is, the screen is a message to me that my writing is not enough, not enough for *it*. For what? It is not simply that it is not enough for the screen. No, not just for the screen. *It* shows my errors, shows my failure. What is *it*, and not enough for *what*? Not enough for the beyond, for that other that saturates the other space, not just as a gaze over the images that appear, but saturates it as a being, as a quasi-ontological condition that is in everything that is there, on the other side of the screen. This is *it*, the gooey thickness of the great beyond of digital writing, a soup where bits of my writing are caught, never quite mine, never quite anyone's.

You must recognize the oddness of this beyond and this gooey being, tied to my opening remarks, to the oddness of the "I wrote this," an oddness that leads me to be stuck on and held by the crime of writing.

In the great beyond, on the other side of the screen, objects exist that are well-formed and final. What is the *correctness*, what is this finality? Word processing *autocorrect* functions are just the most explicit way of telling me of correctness: they tell me that my writing was not adequate.[2] Microsoft Word tells me in a corporate voice, in the delegated words of Bill Gates, tells me that I cannot write, that I cannot spell correctly. I want you to recognize the intense disturbance and ontological oddness of Microsoft Word autocorrect. I may *will* my fingers to move, but I do so in order to satisfy the screen. The right words are the ones that satisfy the screen; in fact, whatever words satisfy the screen are the right words. No matter what my writing, it is only for this satisfaction that I write. The end of every text is to meet the desire of the great beyond. Is this not the work of the subject of electronic literature?

I try and try to give the screen what it wants. Yet there are always more errors in my typing. If my typing is error free, it is because I exactly meet the screen's demand. But it is never error free. The work, the writing I write, will always fall short. We cannot imagine an error-free digital writing because the most perfectly written text is merely one that is allowed to exist by the screen, a text that is allowed to deny its own being in error to meet the demand that it becomes.

To be clear, there are the demands to type specific words: passwords and logins, as I discuss later; or status updates. I must enter the string that is my password in order to receive my email, I must tell Facebook what I am feeling, I must respond to its query, I must answer what's on my mind, must I not? The screen demands I type certain words, in certain sequences, at certain times. But there is also the underlying demand of the other, a demand at work in all these words, an *absolute demand* that I *must type* and must provide words, must make the words survive in the great beyond. The demand drives my activity. Such a demand is in every screen as writing that will be visible when I write. My writing is part of the becoming-visible of this demand of the great beyond, though it is never precisely the demand itself that I write. I write and I am a writer because of this demand, but I can never write the demand itself, never write the text demanded by the beyond. It is a writing that remains latent, and that I must make appear and that I must read. The other's body in writing is there already for me to discover, or at least this is what the screen demands of me.[3]

The term "well-formed" refers to a finite sequence of symbols etc. The term refers to a finite sequence of symbols that is syntactically determined by a formal language. Well-formed has application in linguistics and more generally in mathematical logic. It is also applied to web pages. Well-formed xml or html is properly nested, properly terminated, and so on. The term can also apply to web pages as a whole: a well-formed document is composed of well-formed xml and adheres to the syntactical rules set out by the World Wide Web Consortium and other such organizations. A well-formed document will validate when run through an external application or validator: it meets the demand set down for its well-formedness.

As in linguistics, where an utterance may not be grammatical and therefore not well-formed, a document can be on the web without being well-formed. Well-formed is a standard but not a condition of existence. As in linguistics, where utterances exist in a social field, web pages exist in a field where the well-formed is simply one extreme. Well-formed documents are a class that assumes propriety and obedience, and assumes a range of behaviors and presentations by designers and programmers. The appearance of pages on the web and objects on the screen is far wider than this narrowed culture of propriety. It is such sheer appearance in the great beyond that concerns me and renders my subjectivity. The visual evidence of the thing concerns us all. It is what holds us, locates us, and makes us subjects in relation to the screen.

If you read or hear this now, the errors are fixed, the type is well-formed. You see this, do you not? The misspelling—we could say the "creative glitch"—must be *permitted* to appear on screen. In the simplest sense, the writer must be given *read and write* permission by the system; the writer can only write if this file access permission is given. If you read or hear this now and there are errors—you notice them, the misspellings, the typos; or in vocal delivery, the ums and ahs, the pauses, the gestures, the deviations from the written script, and all the rest—then you know that they are intended, not by me but by the screen, by the computer where I have put these letters. Or rather, not intended, they are *allowed*. I know nothing of the great beyond's attention, only that it allows me. The writer at the computer is never allowed intentionality; the writer is only allowed to meet the need of the screen. Is this not why we speak of writers working with Flash or with HTML5 or with other applications? They become writers by meeting the demand of these systems.

Permission is given. On the one hand, it is ambient, everywhere, tied to the basest protocols. On the other hand, permission is specific: it is there in every word and image from another person. Every appearance online is permitted by others in the world.

To be clear, it was there before. The screen was there on the page, there in print; there was always a screen that passively demanded my writing. I *introject* the screen's passivity. It allows me to pass, permits me to write, and allows my writing to exist. My writing is passive, guided by the network, given to its needs. I am passive, my interior is pacified, my interior ready to offer itself to the screen. The screen commands my fingers.

In this way, I fill the screen, pour myself into it. The network "beyond" the screen is a way of producing a body, producing a knowable body of the subject of electronic literature, producing my body as passive zombied appendage. When literary critics describe "embodiment" in relation to digital writing, they reinforce this zombie shell as our occupation. The network is a way of situating and firming up, of hardening my body. My interiority expends itself in writing to the other. The surface is filled. *Whatever fills it is interiority.*

The zombie eye must see the images. My eye must not be caught on appearance, on the planes of color, on the faint flicker of the screen refresh rate, at 72 Hz per second or higher. The zombie finger must point and click, must type. My fingers must not linger on the keys, must not—cannot—attempt a character others than those given. Again: we cannot conceive a character other than the given. The glitch is inconceivable and excluded. The notion of glitch and glitch art is nothing more than a way of talking about the intention allowed me by the great beyond. The glitch is a play of intention. I can follow a link or choose not to, but the decision is binary and absolute.

The network is an *imperative* operating through the screen. The "must" is the imperative to see. It is the imperative to be an organ, to be a body part attached to the screen, to be a terminal of the network.

Portal or mirror? If the screen is a mirror, my image is elsewhere, teleported into the screen. I dwell on the other side. In this undead phenomenology, the imagined other operates on me through the screen. Writing is a blockage to expenditure. As a delusion, as a hallucination that the other exists, that my words will live on in the great beyond. I write to appease the other's desire. I write to give to and fulfill the other. I decrypt the other's desire through codes. I learn the commands of the application, the menus, and the scripts. I am desperate: to achieve the correct commands, to achieve the commands given to me. I will do as you command, in your name, in all your names, you who give me permission and allow me to write: oh Adobe, oh Microsoft, oh Apple, all the names in your command. In doing so, I give myself over to exactly what you demand.

Weapon body

My fingers are on the keyboard. I type by pressing the keys. Letters appear and string into words on the screen, above and beyond my fingers. What kind of body do I need for this? An able one, with the capability of pressing the keys with consistency and force. Of course, there are large keyboards, left hand keyboards, ergonomic keyboards, adaptive keyboards, and so on. You can marvel with me at the wonder of such assistive and prosthetic supplementation! It allows more and more of us to create the right pressure in the right sequence that produces writing on the screen.

What body are we all trying to be? What is the consistency of that body? It is exactly *consistent*. No particular body but one that presents itself consistently and regularly. This is the biotechnology of waste management, the cyberpower of managing the production of writing as waste.

We may think of what happens on the screen as immaterial and virtual, but the body I am describing here, the body we are asked to present to the keyboard, is not at all immaterial and virtual. It is heavy, thing-like, and clunky.

Why heavy? The weight is precise and measured—it is exactly the heaviness that I apply to the keyboard. Why thing-like? I could press with a pencil, with my nose, with the corner of a book, with my penis, with many body parts—it doesn't matter. Yes, of course, the reference to my genital pressed in the text, on the keyboard, is provocation to say that the hidden body parts, the libidinal investment, is there but is also passified and neutralized. The network is introjected, as I already stated. All the porn of the net is neutralized and neutered. The body terminates at the terminal and becomes nothing but decay and dust. The segmented weapon-body uses any part at all. The body I offer must become *just* this thing with this much weight and the right size to hit the key. Size does matter, as they say.

For all this, the consistent body is clunky: too much weight and there are tooooooo mannny lettersss, as we all know when we lean too hard on a keyboard or dangle our wrist on the space button. Or *clunky* if my fingers are too big, too thing-like on the keys, hitting more than one, and there are wropgn oop wrong letter.

My typing fingers are tooled, precision-made, and weaponized. The military pilot with first-look first-kill cockpit capability or the soldier with a DARPA-enhanced HUD both acquires their targets with fingers and eyes. Is it different for me at the keyboard? For the letters to appear and for the mouse to click, I must move at the correct speed and apply the correct force. I must use a control over myself that fits my organs and my motions into the technical system. No surprise that the same technologists and the same vector of development feed both the military cockpit and the business keyboard.

My fingers are weaponized because they become shells, carapaces for interiority. I may type my "inner feelings" but I do so only if the sensation in my fingers is quelled and numbed, if the carpal tunnel syndrome is under control, if the itch on my palm does not distract too much.

To type on the keyboard is to make myself a weapon. We like to say that there is "embodiment" at the interface. Can we say the weapon is embodied, that there is an embodiment behind or beyond, withdrawn from the weapon? The weapon is the withdrawal of my being.

I must feel the keys but just enough to know that I hit them, and I must deaden my eyes just enough to read the screen. My fingers do not desire to type, they want nothing. They are stubs that strike and hit, but at the order of what? Do they type because of my "intentionality," in a good old phenomenological sense? Do I type following my will and desire to write? Yes and no. Yes, because clearly I select and direct my fingers' actions. But no, because I cannot type *just any action*. I must fit to the keyboard. *Literature, in this sense, is forbidden in digital writing.* The possibility to *write anything at all* is a fundamental condition of literature—I assert this categorically—and precisely this possibility is forbidden to the fleshy weapon-body at the keyboard. In this sense, all that passes under the name "electronic literature" is really typing practice for militarized weapon-subjects. The weapon does not strike the screen but is material or *matériel*, that old sense of equipment in the military supply chain.

And yet impossibly, nonsensically, the result is that all digital writing turns literary, all writing turns or collapses within itself, all of it follows a procedure that re-writes everything from the first.

Fingers are not organs but they become organs. They become the place of sensation. Organs are flowers of libido. I fill my fingers with myself in order to interact with the computer. My squishy body is in relation to withdrawn interiority *and* to the network. The weapon body is hardened; it is entirely at the beckon of the network. I type in the correct way. I look at the correct images.[4]

My eyes are no different. The ocular flesh is numbed and the gaze is locked and loaded. To look at an image on the screen requires that I direct my eyes at a particular zone. Prior to thinking on the specifics of programs such as Photoshop, we must understand the constitution of zones of imagery. To design and author multimedia—here I am invoking the entire field of web and new media development—is nothing other than the creation of this site of vision, this phenomenal zone of the lure and the attraction of sight. It is about placement of the image. The image solicits my eye before I even look at it. When I see it, the image is already in its place, it is already an image to be seen. One condition of looking at a screen is that I know it is an image and not a painting or a random spread of colors. My eye is wired and directed, turned left and right. Even more, my eyes are frozen, locked into a channel. The muscles are numbed. To sit at the computer is to be ready for surgery. Even more, the computer display is an ocular surgery. If you want proof, read the warnings about eye damage from extended staring at the computer screen. All my flesh is ready for targeting and launching missiles, or ready for the twitchy finger in a first person shooter game as the training finger for squeezing the trigger of a gun. Targets include a key on the keyboard, a field on the screen, a button on the mouse, a word to click. The computer screen is a field of targets for my weaponized body. My body is metallic in its response. It is a shell that contains a scuttling and pulsing organism.

Speed is crucial here. Matter at speed is not necessarily fast but it is conditioned to move, conditioned to operate at certain speeds and to operate differently at other speeds. It is motorized. At times my fingers move slowly, at other times fast; my eye may rapidly scout the screen or may deeply focus on an image. I am trained and conditioned to twitch and move at speed.

Zombie computers are compromised and controlled remotely by hackers. Your computer is hacked and runs on a botnet. You do not even realize it: your computer is a zombie. You do not realize it. This is the condition of digital writing. A zombie without knowing it: this is the mindless part of being a zombie. Your computer is carrying out attacks all over the world. We are all on zombie botnets. How could we know? Zombie flesh is a rule of the web. Bodies plugged into the web are zombie bodies.

Zombie materiality is segmented and dismembered body parts—fingers, eyes, and so on—that are tele-directed by a problematic and withdrawn subjectivity. It is the super-production of inscriptions that are "elsewhere," resituated in file storage or on the cloud or simply *on the screen*, where all these terms are metonymic redescriptions of the sending or disturbance of the scene of writing. Zombie materiality is empty of thought, empty of organism. It is disembodied in a precise way. The artist Stelarc announces, "We live in a time when flesh is circulating and bodies are being detached from some bodies and relocated in other bodies."[5] My body is weaponized by conditioning of movements and flows of response. The web must function for *any* body that sticks its fingers on the keys, for any mound of animate flesh that puts its eyes to the screen. Fingers and eyes are anonymous. The web *must* function in this way, with anonymity, without regard for the body attached. Zombie materiality feeds the web and maintains it with life, with a channeling of our life, precisely by this emptying out of any particularity of the organism involved, by this voiding of interiority, leaving nothing but weaponized flesh. The web zombifies my body and in doing so keeps itself alive. My voided interior feeds and sustains the web.

I am speaking here from the biopolitical position of a subject constituted by the life-sucking negativity of the screen and the positivization of zombie flesh. Constituted and positioned, located in this body in relation to the screen.

Is this not exactly the point of electronic literature as studied by literary critics? Is this not the point of digital writing as practiced by all of us? Is this not the point of "human computer interface design" or "human factors" or any of these concepts? Richard A. Bolt's early work on the "human interface" focused on making voice and gesture interchange. The notion was that interactivity would be keyed to this successfully shared modality. But the body at the keyboard is *no one*, it is no subject, it is a zombie. The human in the formula is a generic body. The screen starts from this imperative issued by the anonymous other. The genre may be generously expanded to include left handed people or those with low-vision or various forms of disability. Of course, this simply means that zombie status is available to all. Websites are designed for *any body* to operate. Weaponized fingers mean nothing else.

I am filled with the anonymous being of the network. Yet, impossibly, I imagine through the screen to you. I address you and beg your response.

Crust

The surface of the organism is a kind of "crust," according to Freud, baked solid by exposure to the world. The surface becomes "inorganic," a "special envelope or membrane resistant to stimuli." Beneath this, Freud posits a zone of unfixed, decathected intensities. It all comes down to how we expend intensities. Where is that crust? Where do the intensities surface?[6]

But the surface is the screen, the screen is surface. The other is only a figure traced on the surface. The other is an invention. There is no other and no demand of the other. Better to say, the other is a story told by electronic literature, a story written on what matters: on the segments of my body. It is the interruption, the final segmentation, and the cut that lets me *let it all out*. Nothing is left. Hit the keys and it is gone. It, what is it, it is what is *gone*. Energy, interiority, now expended.

The other is dead and so too my zombie body is necrotic, failing, decaying. There is no demand of the screen. I am no weapon, I laugh at protocols. I can do what I will, expend my energy to create the *waste that is writing*.

The word "necrotic" names dead flesh, yet all living flesh is dying, and thus all living flesh is necrotic. All living tissue is dying through *apoptosis*, through inevitable cell death and sloughing away. Through *blebbbing* and *decay*. In necrosis, the organism is traumatized. It dies beyond. It undergoes a second death, a post-death. Why beyond? The image, the code, the trace in the great beyond is dying as well, is decaying, and is necrotic. I share in the death of the image, in it's wearing away. I share in the decay that we both are (you and I).

The organism is no longer the circle of thought and movement but is encircled by necrotic cells and membranes. There are stimulations and palpitations of the necrotic flesh. It is dying but there is effervescence, excrescence of the flesh, and super-production of the body. Death exceeds itself.

And there is writing, no end of writing.

I pour everything into the keyboard and explode onto the flat darkness of the screen. It does not matter what is on the screen. Seriously: the screen is no matter. Oh, sure there is the materiality of the glass and chrome, but who believes that the image lures us? Matter stops at the keyboard and surface of the screen. The imaginary structure of the on-screen—that great beyond that I entertained earlier in part to construct and tear apart, in part simply for entertainment—requires too much psychology, too much investment in the other. I refuse it: I dismiss the other; I disinvest from all alterity to leave only the great distended edge or rim. With this, I refuse all communication and all messages. I send writing elsewhere, into the break, into the great beyond. It does nothing but expend itself, nothing but produce itself on the surface of my fingers, on the light that hits my eyes, on the waves of sound washing across my ears.

Is this not writing itself?

My eyes burst on the screen; my eyes are blind to anything beyond the screen. What is this blindness? It is the eye that lingers on the color planes of the screen and not on the image. It is the eye that wanders across the glyphs of the screen as across a cave wall, a wandering eye that does not see characters, does not see text, does not see the network at work.

Is this blindness not there in every screen, in every time we look, in every image? It is the blindness of the other, impossibly distant in the great beyond. Is this blindness not the sight of an image that burns the eye, perhaps the images of Abu Graib posted on the net? Is this not the image that makes the body spasm, is this not the pornographic image but the naked, the specific naked other through the screen?

My fingers are full of me. Full of fluid, bone, flesh. My eye is stimulated, tickled, irritated. My vision expends itself on the screen. My fingers expend themselves on the keyboard. This is the terminus of sight and body. I am not a weapon; I am a lover of literature. The blindness I write of, the waste: this is literature, this is what we turn toward and love.

Writing is character, writing is formed, writing is crusted. Writing, oh the pleasures of it, oh the absurdity. Writing is waste. You can talk of the message. You can create diagrams of written communication. Such maneuvers are diversions; blockages that re-divert waste to others paths. Literature is glorious, achieving nothing but the extravagance of expenditure, nothing but the accumulation of written garbage.

To speak of writing, not of this or that writing but of writing itself, is to speak of a spasm or explosion. *The proper name of writing,* what the word designates, what the word "writing" means, is this explosive and spasmic excess, this overwhelming mass of waste, this production of slag and detritus. I ask you to recognize this. You do not read writing; you cannot take in the mass of texts in the world. You cannot take it. The writings exceed you, they overwhelm you, and they bury you. You might write this text, or write that text, but you know nothing of writing, nothing of writing itself. No, our entire species is devoted to producing greater and greater explosive spasms of overwhelming printed matter. Is this not the network? Is this not the web? Not texts, not writing to be read, but writing as massed marked detritus.

Perhaps this spasm is my only point. As I began, it is the only thing. Faced with the waste of writing, the body abjects and curdles. We find a story for this explosion, for this convulsion. We enjoy it with others, with other lovers of literature. We collapse inside when faced with the mass of writing that surrounds us. To read, what the word "reading" means, *the proper name of reading,* is this turning and scintillation of our bodies—scintillation meaning oscillation, displacement, flicker, at the same time auratic and painful and glorious—this is reading as we face the junk heap of writing. To read and to write means nothing else. The excess of writing touches our body. We are maddened pathetic hoarders holding books and papers.

Curdled and abjected inside: it is here that I touch you and feel your touch.

Today this mass is found in microprocessors and computers. On any given day the amount of energy used to power Facebook pages, recipe searches, news sites, and all that the Internet entails is well over 20 or 30 Gigawatts. In terms of greenhouse gas emissions, power consumed daily by the Internet exceeds the expenditure of the airline industry. Rather than burn fuel and fly the earth, the net burns the energy of writing bodies to spew text across screens.

In the dialogue *Parmenides,* Socrates claimed mud and hair and the like, that is, the raw and base heterogeneous stuff, do not have forms. They are "undignified," in his words:

> visible things like these are such as they appear to us, and I am afraid that there would be an absurdity in assuming any idea of them, although I sometimes get disturbed, and begin to think

that there is nothing without an idea; but then again, when I have taken up this position, I run away, because I am afraid that I may fall into a bottomless pit of nonsense, and perish.[7]

In the dialogue, Parmenides himself suggests that with age and with wisdom, Socrates will include these within philosophy. The waste, the mass of mud and hair and dirt, the mass of nonsense are what we face on the screen. Writing a philosophy of waste.

I send you an email. It does not matter the message, it is the sending that surfaces energy and produces the event. The message is gone. I may await its return. Like a fishhook in my skin, I await the reply. The stimulation, the irritation to my body in sending is what matters, is matter. To say it is about your reply, to say it is about the other's word, is to jump too far ahead, is to believe too deeply in the structures of communication, in the possibility of the phatic nature of my email. I hit send to waste myself, to get wasted. I am inorganic crust covered with words, words existing in the waste piles of screens and books.

We, who are we? All those "on" the computer. All those at the keyboard. All those on the Internet. We are artists. Our writing is an invention of explosion and vanishing. Literature is what remains: it is open, open to all; we can take hold of this writings.

The interface is an explosion. I write to bring about this explosion.

II

For example

A page from Jake Chapman's *Meatphysics* or "Lip" by Alan Sondheim and Kim McGlynn. Examples of what?

"The sun deluges its chemical holocaust to the earth with blinding indifference" begins *Meatphysics* and the tide follows for hundreds of unnumbered pages of the "vagrantly purposeless" (back cover).[1] Information disgorges and dissipates. Writing absorbs and saturates. Fast-forward flow. Nothing can be said about this book. Is it understandable? Is there even confusion for the reader— can we touch it: your confusion, my confusion?—which assumes some reading and some hermeneutic at work? Is the source of understanding in the code that eats at the writing, described by the book's publisher as "a file-corrupting literature-machine that contracts and expands without final cause"?[2] The code spews over the pages. Is there a process or algorithm at work? I know you are thinking this: can this book be called "codework" and fit in a genre and practice that applies code processes to text? Does this genre even exist, which I doubt? (Just as I doubt all genres of literature.) Can we discover the pathology at work here, the underlying complex? Can we analyze the text and distinguish machine from author? Can we read the clever complex of the author in the wash of writing? Can we presuppose a semiotic that bookends and references the partially sourced emissions at work in *Meatphysics*?

How about this: Jake Chapman, of the brothers known for their "fuckface" visual artworks, writes the book and signs it as the author Jake Chapman. Or, Jake Chapman writes the book and signs it as the author Jake Chapman, and a computer program alters it. Or, a computer program writes the book and signs it as the author Jake Chapman. The alternatives articulate the economies of writing the book. We are always in the midst of these economies: of the authorial project of Jake Chapman, of the brothers known for their "fuckface" visual artworks, or the hybrid economy of the human–computer collaboration, the "intermediation" recently described by Katherine Hayles.[3] You want to know the answer, surely you do? Sandy—you tell me—you idiot, complete the analysis and put the book where it belongs, extract the machine and the psyche, and lay them bare. What is this guy thinking? What is this code doing? Please, you ask, make this a writing that economizes rather than generalizes.

Here's the answer: Jake Chapman, of the brothers known for their "fuckface" visual artworks, wants to shock and disturb, and the book shocks and disturbs. No, that's not the answer.

"Lip" is from a 1995 Ytalk conversation between Sondheim and Kim McGlynn. "Extreme sexuality" is how Sondheim describes "Lip." The writing is straight from the screens that McGlynn and Sondheim stared at. It begins "erection...around my balls...yes...across the screens...yes..." and continues through the repetitious and intense writing of the netsex between Sondheim and McGlynn.[4]

Extreme sexuality is not because of the content but because this writing is the extremity of bodily sexuality, extremity as surfacing or as distensions in and of a surface. Ytalk was a variant of the UNIX "talk" program allowing real time chat conversation. It was similar to IRC or instant messaging but with significant differences. First, Ytalk does not maintain the order of communication, but allows messages to overlap and insert. Second, it transmits the message character by character, rather than message string by message string, as in instant messaging. The results are lag and acceleration, hesitancy and withholding, force and exhalation, flows and stutters, streams of characters from unnamed talkers, keystrokes and erasures, pauses and accelerations of typing fingers.

The specifics of the interface organize the libidinal economy. To be specific, not so-called and never-occurring "materiality of media," where an idealized media form determines content, but a sieving of contents that libidinize and inhabit the media form. I leave the phrase "materiality of media" unsourced on purpose, to invoke the arguments associated with Friedrich Kittler without necessarily asserting that they are his arguments. Psychoanalysis provides terminology for the results. The splitting of the terminal screen between the talkers makes the interface an ontological rupture and haunting. The interiorized topography of condensate and coagulate and chewed-over ASCII characters is dominated by the imaginary. The introjected screen becomes a site where the talkers are projected, the unnamed talkers, perhaps "I" and "you," projected and organized in the service of desire. What is given to the screen? Body parts, secretions, pressures, breathes, all excessively given.

The writing ends " ... i can lick it taste it ... taste itla;LKJ lip." The end or limit I observe is pre-symbolic and intimate, that is, literally interior and inarticulate. The extreme again, the edge where ASCII turns organic. Do I read the limit, do I touch it? The conclusion of "Lip," as well as every character and spacing of the writing, is not a sign but an organic membrane. Pre-symbolic in a specific manner: imaginary at work from the first, from and with the body toward an other who is always absent. The tethering of body to the screen imaginary of digital objects points to the relation Yukio Mishima explored between flesh and the ideality of words. This imaginary is keyed to the organic turn of the buffered characters in the talk session. The lags and delays are latent here: reading as the latency of breathing, of secretions.

Of course, there are institutions at work here. "Lip" and *Meatphysics* are signed and published and delivered. You can understand everything here as "netsex" or as indicative of the puerility of the author, as if to say "oh, typical of Sondheim" or for that deranged Jake Chapman, of the brothers known for their "fuckface" visual artworks. Make a case for it and out of it, but cases of what, examples of what?

What if these pathological cases—in all their flatness and printedness—are the best and perhaps only examples of electronic literature? Works typically described as electronic literature—or other cognate forms such as digital poetry—are not, or at least not for the reasons claimed. Or more precisely, those works may be digital electronic literature (etc.) but the qualities typically emphasized by critics in those works are not qualities of electronic literature, but are instead part of the phantasmatic role literature plays for criticism. Electronic literature means the literature of the electronic, or—I prefer—of the net. Not the problem of defining electronic or digital or online literature but the problem of writing the net as the problem of literature.

Framing this is the historical relation to the book. First, to specific books and specific protocols, all as families of writings; second, to the total book, in the sense that the book and the world adhere and collapse.

The concept of *file* and *format* are conceptually radical in ways that we do not grasp. A format is the format of a file, which is to say that a format is a format, and a file is a file. Consider RFC #1341 from 1992, entitled "MIME (Multipurpose Internet Mail Extensions): Mechanisms for Specifying and Describing the Format of Internet Message Bodies," authored by N. Borenstein and N. Freed. In the simplest sense, this document defines the format of email messages and their attachments. Several revised RFCs follow and obsolete this one.[5] Notably, RFC #2045 and #2046 and later #4288 and #4289 expand the mechanisms for specifying and describing to "arbitrarily labeled content." The format of Internet messages was defined earlier, in 1982 by RFC #822, but, as later RFCs frequently remark, little was done to define the message body. The omission was precise: the body was undefined and could be *anything* that could be carried by the ASCII wrapper of the message header. As the later RFCs note, the goal of the MIME type mechanisms is to provide definitions for "textual message bodies in character sets other than US-ASCII" and "an extensible set of different formats for non-textual message bodies." The goal seems very simple but MIME type is in fact one of the most ungraspable and complex aspects of the Internet. The extensible set makes it possible for the message header to refer to any format as content, where the notion of "any format" remains undefined and open.

While the specifications in the RFC document are directed at email messages, they subsequently were expanded to apply to protocols such as HTTP (for all webpages), RTP (used for VoIP), and SIP (used for streaming video). In short, MIME type identifies and names the format—the type—of any and all files on the Internet. IANA, the Internet Assigned Names Authority, references the MIME types when it lists allowable media. These include application, audio, example, image, message, model, multipart, text, and video. The experimental types, or the example and model types, mixed in with the more specific types, point to the axiomatic function of MIME. More precisely, as a set of *mechanisms*, the MIME type specification is a *machine* that writes the name of *any* object and includes it within the net.

The World Wide Web Consortium declares the following:

> …the architecture of the web depends on applications making dispatching and security decisions for resources based on their Internet Media Types and other MIME headers. It is a serious error for the response body to be inconsistent with assertions made about it by the MIME headers.[6]

In truth, nothing prevents such dangerous assertions; nothing requires a file to be serious and dependable. The looseness of designation, the fact that a file can be other than what it claims to be allows for spoofing and malicious attacks through hidden scripts and applications.

What is a "format" or, for that matter, what is a "file"? File names are introduced around 1962 for purposes of identification. The fact that these problematic terms are described in terms of mime, in terms of mimesis, situates format and file in a history of similitudes and mimetics. It is possible to say that a file has a format and that a special type of file defines formats. It is possible to refuse the question and say that files and formats are definitions of data, which is to say that they are categorical and indefinable types, or better that they are mechanisms and axioms of media.

What does MIME type mime? What similitude and mimetics? The fact is that we are unable to define what we mean by a file name, other than to recognize it as yet another form of debt or obligation to the zero degree of the surface of media. The author gives the work a name. File names are one obligation between the writer and machine. For a start, the obligation that the name of the work names the work. We honor and invest in this obligation when we create archives to preserve work, or when we create an entry for and about a work in a database such as ELMCIP Knowledgebase or the Electronic Literature Directory. The name contracts that the work exists and will be maintained through the persistence or hardness of the link between name and work. When we are concerned with toolchains and software upgrades, it is not just the operational question of whether we can access Michael Joyce's *Afternoon* on the old Hypercard files—or more recently whether we can recreate Flash files in HTML 5—no, these concerns with the archive and the material are about maintaining accounts and obligations of and toward the work. We are never simply considering the materiality and specificity of the medium; we are always tallying up our debts.

oooo ooooooooo

The o in ROTFL, or OMG LOL and TTYL LOL, the first is "o my god laugh out loud," the second "talk to you later lots of love," where you deliriously read the l as love or laughter through horizons of relevance caught in your body and that other body texting you. Or the o in LOLZ, where only a noob with two o's would read lots of love. The o in I/O where the release of output is opposed to intake, introjection, and to the "I" of enunciation. The o of fort/da where Freud explained the fort as the exorbitant drive to display and exceed the boundaries of self. The o of net sex, where the libidinized o or oooo is readable as the height of pleasure, as moan and orgasm, as dissipation and lassitude, or possibly and unknowably, as just a heavy finger on the keyboard. Combine o with colon and dash, and you have a face, a readable emoticon, but of what emotion? Is the o wonderment or is it a big yawn? Every emoticon is an avatar body, the o of the other, big and small.

But also the o of apostrophe, that rhetorical figure of literally "turning away" or breaking the flow of discourse: "O death where is thy sting," "O my friends there is no friend." Speech is directed to an imagined person, an abstract quality, or to an absent other. Apostrophe is a presence summoned through discursive rupture, narrativization and intense emotion.

For that matter, consider the typeset apostrophe, ASCII character 39, used for quotations marks or to delimit string constants. In either case, a container or siting of an utterance in language or code, but also a character used to voice speech itself, or as an acute accent, or sometimes as a glottal stop.

EE EEE ETEE ET EET TTE

The beginning of a short text by Alan Sondheim. It started from a "prose poem of hysterical politics." The first line of that beginning was "I slaughtered hundreds, I can't believe it now, war is hell" and it continued, concluding "I can't speak it my mouth stuffed." The writing is confrontational, anxious, provocative, wailing, and pleading, typical of Sondheim's codework poetics. He writes a performative machine that troubles relations between reading and uncanny reference, and in doing so, sets a standard for the literary in electronic literature. He explains that the initial or source writing was:

> translated into Morse code using the Morse program in /user/bin/ games. The Morse code was then re-justified to fit a 72-character ASCII line. This line was then hard-broken using the fold -w command. The command began with the lengths of the 18 characters, down to the length of a single character (which in Morse could only be "space," or T or E). If the length, for example was five characters, the original line would then be decoded back from Morse into standard English; because of the line breaks, the words—the very structure—of the lines—underwent transformation. The final result was a series of stanzas reflecting the broken code, as if there were a kind of interference—which there was—as if the interference were a structured line-noise, for example.[7]

EEEE T E ETE E TEE and so on for lines and lines

"I slaughtered millions" and "I can't speak my mouth is stuffed" is transfigured and processed. These phrases speak of voicing and enunciation of the subject, as well as violence, suffering, bondage, transgression, bodily investment, ritualization of writing, and problematic boundaries or bindings. EEE EET TE TEE ETE E TEE EEE. There is a kind of speechlessness. What speaks in the E and T and spacing of the text, or for that matter in the o I spoke of already? The sublimely literary return of bodily investment in the codeworked text is fantasmatic and imaginary.

Of course, the source text and the transformation are readable in a more or less straightforward way as avant-garde writing practice. All a question of the "right aggregate," as Samuel Beckett put it in *Stories and Texts for Nothing*.[8] The text is a supplement produced through technical processes, and the fantasmatic voicing is the articulation of exactly this supplementarity. Such a reading finds a second text in the haze of processes, a text discerned through knowledge of the codework involved. The desire for origins, for knowing the code and organizing and organicizing the text, or for reading the object o, puts language in its place, both in terms of absent bodies and presented texts. In short, it involves rhetoric.

The opacity of net writing is the executable code of generalized metaphoricity, a figural schema that makes even the most hashed up bitstream readable. It is the inheritance and survival of rhetoric, from the enigma of the E or T or o, to the most complexly coded application or web page, to the simplest spam email with the uncomfortable address that hails me as "my dearest" and begs for support or offers suddenly discovered wealth. Every reading of net writing moves from disorder, one rhetorical qualification of data, to arrangement, another kind of data. Disorder as data is any code, any message. Arrangement adds reading protocols to deal with the given-ness of data. Such a movement, however construed, is exactly the definition of rhetoric as "institutio" in the sense of Quintillian's *Institutio Oratoria*: arrangement, siting, voicing, and the creation of what we call institutions, with implications of culture, pedagogy, and power. Such institutions are evident in the emergence of writing and research on code, whether by Google, or in academic branding such as code studies and the like, all premised on the problematic of unreadability and on the rhetorical institution that deals with this unreadable situation. The rhetorical institution makes the unreadability of code an occasion for art and scholarship. Foregrounding such rhetorical operations emphasizes the black box of the subject of the net. The most elementary explanation of the technical processes involved in any net writing highlights artifice, insisting on behaviorist tropisms that produce effects of subjectivity, readability, textuality, and the like, presuming latent performativity for every protocol, and presuming an event for every inscription on the net.

Net writings offer images of otherness. The subject is marginalized: nothing more than the inscribed graphic character on the one hand, and the great outside on the other, that wilderness of materiality, of energy and light and particles. My reconstruction of Sondheim's text or of the letter o sutures the subject to the symbol, to the repetition of the mark. Think of a body rewritten through cultures of net inscription, think of this body uttering mournful or ecstatic poetry, think of writing ourselves into and out of and through existence.

OMG LOL

I mean: what performative traces are there in EE EEE ETEE ET EET TTE? Not the mnemotechnic supplement of an event but the writing of oblivion. In place of memory, the shapes and spaces of code poems give the minimal repetition of *writing itself*, writing as a mark—and recall that it is repetition, fort/da, that Freud identifies with the death drive that takes us beyond the self. In such a supplement we can speak firstly of the literal poetic datum as an originary givenness. Secondly, we can speak of communities of writing in relation to the poetic literalness of the mark carried in net writing.

As always, rhetoric explains what the poem already states. The very dissolution and absenting of the subject creates ecstatic and exotic voicing in the E and T and spacing and o, and in the entire characterology of the net. The basic protocol that assumes the ornamentation of metaphor and the empire of rhetoric makes the thick semantics and relevances of the subject withdraw behind technical functions. If you read and decode embodiment in media, it is precisely because you presume yourself subject of and subject to inscription. As such, the subject is literally sublime. There is writing on the net and authors and readers are always elsewhere. The classical notion of the sublime inherited from Longinus is poetry as ontological utterance of being always elsewhere. Instead, the literalness of codework poetry emphasizes invention and extension of being as experiment.

You need to think in terms of the political economy of the net as an inscriptive skein, where "skein" is the morphology of an epigenetic landscape of inter-imbricated organism and information, and "inscriptive" is sites of marking and practices of reading through the imaginary. It is an economy that produces by redefining the inscription of organism as protocols of recognition and address. It is an economy of negation. Heinz von Foerster wrote:

> Since nothing in the environment corresponds to negation, negation as well as all logical particles (inclusion, alternation, implication, etc.) must arise within the organism as a consequence of perceiving the relation of itself with respect to the environment.[9]

Consider system lag. You know what I mean, we all experience it. I type a text to you and send it, but it lingers in the outbox. The reason may be only a few bars of AT&T 3G service, or poor load management, or bad hops across the net, whatever, but the message is slow, with no report coming in, your response lost. I am held, waiting for your text. It could be the same with an email or a form on the web. Held in place, bound, speechless, all protocols canceled, subjectivity disinvested. Yet soon the slightest screen flicker bodes your response. The screen freezes: these characters irradiate with waiting and absent voices. Such a limit is the literalization of the sublime, literally beyond the senses.

Even visibility breaks down. This collapse of vision is very different from the semi-ironic knowledge of the constructedness of vision in contemporary media culture. By this I mean our habituated argument, the commonplace of theory that relativizes the visualized object. The result is a kind of sense certainty not of the object—none of us believe in the simple reality or mimetic presence of the object on screen, we are all able to critique the resolution and construction and Photoshop potential of images— but rather it reinforces for us the certainty of the representational schema that ties the thing to data. The falseness or inaccuracy of the depiction, the fact that we continually apologize for our images or explain that they are possibilities, visualizations in the plural, all this constitutes ways of insisting on epistemological control of the schema: we recognize its fictionality and in doing so ensure that it produces an institutional truth just as it shows the falseness of images.

Something quite different happens in the graphic becoming-literary of the literal. The character is inhabited by otherness, by a fluidity of bodies, all construed through the mark. The very well-formed nature of the graphic, its presence as symbol, and its precise Unicode and font rendering dissolve through the imaginary, to become an impossible presence. It is this that I call the literal poem.

WTF. Here is my apostrophe. TTYL LOL.

Leet or 1337

Consider the pre-broken net domain of leet or 1337. Often described as a "corruption" of written text, the practice of "leetspeak" or

"13375p34k"—and also known by many other names, such as "hacker" or "hackerese"—emerged in BBSs, listservs, and MOOs, and is characterized by replacement of alphabetic characters by similar appearing numerals, but may also involve other replacements or radical abbreviations. Unlike some related argots or jargons such as hexspeak, leet is not a strict one-to-one encoding. There is no standard for the practice of leet; rather, its diversity varies with the protocols and group cultures of online writing environments where it appears. The f/ph replacement in "fone phreaks" is early leet, as is the s/z replacement in "warez." The character "a" can become "@" or "4." Either substitution is based on optical recognition of characters and transposition in terms of visual similarity. One leet usage is "teh," a mis-spelling of "the." This is a common typographical error and automatically corrected by word processing software such as Microsoft Word. On a standard QWERTY keyboard, adjacent fingers on the left hand type the letters T and E, and H is typed by the right hand. Coordination of the hands conflicts with the incorporated skill of touch-typing. The auto-correct feature lets you speed past your errors. With "teh," the refusal of proprietary pseudo-intelligent software and the proximity to repetitive and forcible bodily rhythm intensify "the," so that "that is teh lame" means "that is the lamest." Similarly, the fact that the 1 and ! characters are on the same keyboard key leads to a practice of swapping and repeating for emphasis: "y0 d00d th1s 5h1zZ47 R0Xx0rzZ!!!!!11."

Leet is character encoding for eyes that see and a body that experiences. Leet is read for the visual glyph, but also read through the glyph to the virtual or invisible intentional character suspended behind it. Every glyph or graphematic unit is read by rules of encoding and substitution but also through the phenomenology of address, where the glyph is meant for the reader. The illegible carries the legible within it, and leet is a reverie of the digital character. Theorization of leet only demonstrates its false order, a symbolics of elemental husks eaten from within. Leet as abbreviation allows for speed and efficiency in communication. The reduction or simplification of messages is possible within the framework of writing environments such as listservs or chatrooms, in terms of group norms and assumptions. In fact, simplification is a function of this framework, and the speed gained is only possible by the accumulation and implication of group knowledge in the environment. As a derivation of the word "elite," leet presupposes identification with a particular habitus and shared cultural capital. The leet writer is part of a group and identifies with that group in the very act of reading and writing leet. As elite, leet references a sociology of hackers and users, of those "in the know." To read leet is to recognize inclusion in this habitus.

The cost of recognition, however, is enforcement of the protocological constraints laminated on every string of leet. A common use of leet is for undesirable or illegal communication in monitored or censored environments. Leet allows participants in online gaming to swear without being kicked out by monitoring software. A leet formulation such as v1@gr@ can slip by email content filters. Similarly, wares or cracked software becomes "W4R3Z" and "porn" becomes "pr0n." Of course, some sort of filter could flag these terms using regular expressions, but leet as a practice of corruption means writing until there is no possibility of detection and recovery. While a filter might flag four-letter words beginning with "p" and ending with "n" as a high probability of being the word "porn," there is less chance of catching the expression "spl01tz" as leet for "exploits" or hacks.

A related use of leet is obfuscation. In general, code obfuscation is a practice that makes computer programs difficult to read and understand, typically by reducing any text-like formatting or by adding arbitrary formatting. The result compiles and runs on a machine but appears as an unintelligible mess to human eyes. The goal is to conceal information, whether from possible thefts and reverse engineering, or as a means of spamming. There is also a thriving practice of recreational or artistic code obfuscation, such as the Obfuscated PERL Contest and the International Obfuscated C Contest. Leet allows a basic form of obfuscation. For example, leet provides a quick means of generating passwords or user names as unintelligible strings. In online gaming, where administrators or higher-level players can eject players with a simple "!kick username" command, the leet-generated username is a simple encryption, hard to type and easily misrecognized.

Leet is corrupted by the work of overlapping and collapsing digital domains, a palimpsest of intention and regulations condensed anagrammatically. In one direction, leet is formalized as it grows larger than specific communities and becomes a general dialect for net discourse. No longer elite, leet becomes a commercial parlance. Even the rawest newbie can begin using leet. Anyone can set their Google search page to leetspeak, and players on Jeopardy bet $1337 to show that they are in the know.

In the other direction, leet remains problematic. "Corrupt" text implies a cleartext suited for the domain at hand, but clearly leet is a principle of generalized corruption. Let me insist: there is no final mapping or code, neither for the individual characters and words of leet, nor for its practice. Both remain open and are up for grabs. All text is already encrypted in relation to the interminable working of the net. Leet references the absent body and its corruption is the internal processing and subjectivity/introjectivity of readers. This means that every character encoding and every string on the net is potentially leet. 1337=1111=0000.

III

Survivable communication

In 1962, Paul Baran submitted a report entitled "On Distributed Communications Networks" to his employers, the Rand Corporation. Of course, Rand was the think tank behind America's cold war strategies, dispensing advice of mutually assured destruction and other scenarios. Baran's report summarized a series of simulations of the thermonuclear destruction of communications nodes in networks of various configurations. It postulated a science fiction narrative of the post-nuclear survival of communication capability or "survivability." Baran's famous plan for "distributed networks" predicated survivability of communication on dense and redundant super-linking. The centralized network is obviously vulnerable, since destruction of a single central node destroys communication between the end stations. As a result, in practice, Baran writes, "a mixture of star and mesh components is used to form communications networks." Redundancy was the key. Baran defines "survivability" as a "measure of the ability of the surviving stations to operate together as a coherent entity after attack."[1] In short, you design a distributed network full of waste to guarantee that it can communicate beyond the apocalypse. Distributed networks mean nothing other than production of surplus, of noise, and of waste.

The noise of the enemy attack can swamp the network. At the limit, which meant thermonuclear war in 1962, the network is destroyed. Weapons threaten communication. Survival of communication is war by other means. The catastrophe of the limit of communication is where all flesh is stripped away, all bodies and other carriers of semantics and messages are burned to ash. What remains is the network as the yes or no presence of response. The other's response, yes, yes, but the other as subordinate, as part of our side. Redundancy preserves the map of friendly space where response means yes, and where the war continues. Baran repeats over and again: the goal is survivability. A large number of nodes and high interlinking preserve communication even after thermonuclear war. It is vital that you recognize this convergence of the end of human life with the persistence of network communication.

After the initial proposal, Baran followed with a 1964 report containing explicit recommendations for building such a redundant eternal network with the goal of maintaining Command-and-Control after the bombs drop.[2] The recommendations included notions such as packet switching and unreliable connection-oriented transport. Exactly these principles were subsequently implemented by Len Kleinrock and others as a basis of the network of "interface message processors" that became ARPAnet and later the Internet. The well-known notion that the Internet was designed to withstand nuclear attack is of course founded on this origin myth tying its early history to Baran's reports. The myth is widespread—you can Google it, or look in an excellent book such as Alexander Galloway's *Protocol*.[3] Despite this, Kleinrock and others, such as Bob Taylor but including Baran himself, insist that this is a mistake, a kind of after-projection of Baran's thoughts. Kleinrock says he did not draw on Baran's report in creating the Internet. He seemed content with communication that was not built on the dream of survivability.

Kleinrock's student Charley Kline sent the first message ever on the Internet on October 29, 1969. The message was intended to be the word "login," precisely in order to log in—that is, the message itself was an attempt to open communications in order for there to be other messages. Before this word could be transmitted, but after the first two characters were transmitted, the system crashed. As a result, the first Internet message was "lo." No response after the disaster, in Baran's sense, but nonsense, not survivable communication but the interruption of being at the screen.

In fact, scenes of this interruption are everywhere in the computer: they constitute our relation to the machine. They are a philosophy of the somatological scintillations of our bodies as we couple with the network. Consider Alan Turing's foundational paper on "Computable Numbers."[4] It is the interruption of the fullness of human memory by the finitude of computable numbers that necessitates the machine, Turing's finite state computing machine that is in turn the basis of all subsequent computers. When Turing posited the machine that would later become the computer as a device that writes to and reads from an endless strip of paper, the act of computation—remember computability is the central goal of Turing's argument—or the scene of writing, if you will, is modeled on the prosthetic extension and supplementation of human consciousness, which in turn requires the interruption of our body. Turing describes the continuous feed of paper through the machine, stating that "the 'scanned symbol' is the only one of which the machine is, so to speak 'directly aware.'" He adds, "We may compare a man in the process of computing a real number to a machine which is only capable of a finite number of conditions." The gap of the body and the screen is already in place in this comparison. Turing's justification for computable numbers to be calculable by finite means "lies in the fact that human memory is necessarily limited." We deal with a domain of limitation. We deal with mnemotechnics, with technics for memory. The limited and finite aspect of memory is opposed to the non-limitation of human interiority.

Baran's network is not precisely the Internet. It is more like the fantasy of the Internet, the net's fantasy. What is this fantasy? Precisely survivable communication, precisely that some voice lives on after the interruption of being and the wastage that is writing. The catastrophic limit is the basis of communication in the network. The interruption of the body and the explosion at the interface are the conditions for producing the dense waste aggregate that is the Internet. What survives? Utterances, a voice, a response. Surely this is the dream of all writing: the survival of structure, the transfer of voice, and the continuation of being in the technics of writing.

Ping poetics

Only the imagination is real.

William Carlos Williams

Hello Baghdad. Hosts are open, packets receiving. I'm in Iraq in milliseconds. But the DoD turns me back, their firewall refusing to echo, ending my request.

I tell you the net is a field of waves and radiations. It takes us in and we take it in. It is netting, apparatus of capture. The net is defined as always on, connections always open. The earliest Internet Request for Comments, or RFCs, make fragmentation and addressing basic features of the net, not reliability. Questions of accounting and the economics of transfer—for example, this is the source of the term "logging on"—turn to problems of an interminable and intransitive writing. Once in the net, I can speak of datastreams and packets, of addresses and codes, and of world travel.

Here's how I went to Iraq the other day. I rode the net and wrote into the net. Here's what I did: I ran pathping in the command prompt that still remains below the surface of Windows, ran it from my home computer to a.dns-server.iq, the primary name server for the country of Iraq. The text included the word pathping, which is a command invoking a program of the same name, and included an Internet address, which provided the destination for the pathping utility.

What is pathping? Briefly, it combines features of the older traceroute and ping network utilities. Versions of pathping or these other programs are built into most computer operating systems. The traceroute feature collects details of the path followed from my home computer to the address I supply on the command line, in this case a.dns-server.iq. Traceroute maps out nodes available on a system, showing paths, times of route, and so on. The utility pings network hosts along the way, requesting an "echo response" indicating that the host is available. Ping measures and reports problems of availability (machine available or not on the net), latency (time of transmission), and reliability. My computer sends out a datagram or a formatted "packet" of data consisting of nothing more than a request for response and "time to live" or TTL value. The latter is an upper limit on the number of host servers the packet can be routed through before reaching its destination. The TTL is designed to prevent the net from being clogged with endlessly circulating packets unable to reach destinations.

What was the outcome? A writing of my being on the net. The result starts with my pathping command.[5] For comparison, I ran another pathping from my home computer to defense.gov, the server for the U.S. Department of Defense.[6]

What is written here? It took seventeen hops from my computer to the Iraq DNS server and took twelve hops from my computer to the Department of Defense server, or at least until the pathping reached a point where the trace was blocked, presumably by the DoD firewall. I can see my route to Iraq passed through several hosts in the USA, including Washington DC and New York, then through several hosts in the UK, and finally to the Internet Protocol address 194.117 reserved for Iraq. I can also read the time taken for the ping echo message to return to my computer from the Iraq address. The result is a topographic measurement: 122 milliseconds from me to Iraq's presence on the net. A topography of immediacy and presence through my machine.

What is the * or asterisk in the text? It is lack of response in answer to my request. My packet requesting a response reaches the DoD but I wait in vain. The time to live is exceeded and the packet dies and disappears from the net. The DoD ping is lost without an echo. My request for a response is never answered.

There's a masquerade to ping. Spammers send out pings to find new hosts, new sites to exploit. Is it not strange that the echo response, "yes I'm here, on the net," is an invitation to infection, to invasion and incorporation into a zombie network a more and more websites refuse to ping. They filter out the response. Back at home, the sender gets asterisks indicating that the message ends there. Of course, this still maps the boundaries. I can still draw the lines of the net through the asterisks, but it is a withdrawn net, a topography of enclosures and fortresses, rather than responses. Or better, we would say that the net moves from one kind of response to another, indicating the basic relations and positions underlying response. The first is display of the site, an opening; the second a closing, a display of the surface and edge.

Traceroute

If you know enough, think of the Traceroute project started almost ten years ago through the trAce Online Writing Center by Alan Sondheim, Sue Thomas, and others, a project that traced a singular image of the net itself. It would be enough to see this project as the exemplary instance of the becoming-literary of the literal, of the nettings and knottings of the net. This is the "unary mark" and limit I am targeting, caught in the constraints of the Internet's "physical layer."

This writing is not—or hardly at all, ever so little—a text. Such a statement is both philosophical and political. Philosophically, I say text is woven toward alterity, writing reading toward others. In order for there to be a text, there must be intersubjective structures of communication. Text requires response. Even if there is no sign of response, this lack is the signifier of the Other's response. The other is written in and through text always in part and in whole my desire to the other. Translate this into any discourse you want—that is, re-frame it in terms of institutions—and we can say that there is always a crowd in the text.[7]

The network is composed of layers, or so we are told. Four layers in the TCP/IP or DoD model (Application, Transport, Network, Data Link), seven layers in the more recent OSI model (adding Presentation, Session, and Physical layers). Lev Manovich describes two layers, cultural and computer, a distinction so large as to be useless. More or less layers, take your pick.

Layers are infinite and their distinctions infinitely paradoxical. Janet Abbate describes the paradoxes as spatial and temporal. Spatially, information received appears as the highest layer, as the logical abstraction of the lower layers. This is the paradox of a net topology contained but invisible in the signs that appear on screen. Temporally, information received appears as the latest layer, as the outcome of processes in earlier layers. This is the paradox of a time difference contained in the sign, a diachronic encryption of the real time of the net. These paradoxes bind layers to aggregations of expression, to certain sign thresholds, to consistencies of performance. Not only this, the concept of layers itself is bound to economy. Layering becomes the effect of capitalization of the apparent throughput from lower to higher and from earlier to later. The thing on screen as the outcome of the net's work is the premise of digital economics.

What takes place? Messages moving across the net as an inhabitation that is named communication. These are objects for me, parts of the "field" of consciousness. All layers are simultaneous in this field. Even the most bare and functional description of the stack assumes regulation between levels, assumes skills and practices of operators controlling the system, and assumes the phenomenologies of communicating through the net. Higher order cybernetics and systems theory includes the organism in communication. The program inscribes the organism through no particular marks but through abstract encodings as iterative, interactive couplings of organism, and environment as the extension of the organism. Organism is organism plus environment, while environment includes organism.

The net is, in this sense, a single abstract machine for inscribing the subject. Winograd and Flores' crucial notion of domain-generating "breakdown" is a way of understanding such inscription.[8] Except, breakdown on the net does not occur at thresholds or limits but is a catastrophe immanent and spread across the whole topology. From the first, the net is a breaking and broken-down system.

Consider TCP, the Transmission Control Protocol, and keep in mind that the network's transmission and application layers are built to deliver. TCP moves packets of information around the net in a single stream of data and commands. TCP is, after all, a reliable protocol, guaranteeing correct delivery, and yet is built via fuzzy logic on the unreliable net. TCP assumes communication, "association between two or more entities, with or without regard to any path between them," in the words of the original protocol document, assumes communication as association, that is, assumes the net as alterity, as the presence of others, as the multitude.[9] The crucial point is that the stream maintains the priority of delivery, queuing up data packets in the receiving node's buffer, and guaranteeing that messages arrive with the right parts and in the right order. TCP is writing that implies a textual model of reading order and hierarchy, of packet segmentation as annotation, and packet length and format as closure of the book. The writing of TCP is a contractual relation. The segments and packets, addresses and check digits, are dedicated and written toward the other. TCP creates virtual circuits between nodes that are listening and ready for association. TCP is a philosophy of alterity, where "I write" means "I listen for the other, I wait for your reply." More than this, TCP assumes that the other will listen and reply. The protocol's reliability is because you and I, you other and me myself, already enter into this listening relationship, in every TCP-driven communication, whether email or html page. My existence online is a contractual relation to listen for you. Such a contract is writing that situates and precipitates the other and myself in the presence of the network. This last is important to the crowd: the network is silent witness to our bond. As a crowd we are segmented, written with the network. We are in our place, receiving the signs on screen.

Urgent interruption

By contrast to TCP, traceroute and ping work with the IP or Internet Protocol, a connection-less best-effort datagram service, with no guarantee of end-to-end reliability. IP and TCP coexist, but TCP can be said to ride on IP in the hierarchy of Internet layers. The IP domain of net packets remains a pack without direction and place, only committed to increase and flow. The relation is unstable, a "leaky abstraction" in the words of Joel Spolsky.[10] Texts written and received against a field of messiness and flux, writing as gesture and display. As defined by Braden, "IP datagrams may arrive at the destination host damaged, duplicated, out of order, or not at all."[11] IP allows the undefined, the infinite, the failed, and the wild. TCP attempts to describe and formalize the underlying spew, to treat it as reliable, to leave a mark in the place of the leakiness of abstraction. The result is any number of problems of formalization many of which are exploitable by hackers through phishing, spoofing, and Denial of Service attacks. For example, a "ping flood" is a simple DNS attack, clogging bandwidth, while the "Ping of Death" exploits the upper limit to the size of a message or text. A host is tricked into re-assembling a message larger than the maximum IP packet size (65,535 bytes), even if it is larger by one byte, leading to a buffer overflow and a crash. Note here a basic poetic problem of length and form of the text, we might say of the form of the book beyond the book in the sense of Edmund Jabès or Maurice Blanchot, here challenged by a kind of production that enters into the form and destroys it, with comparison to a range of innovative poetics.

Every packet of TCP data contains an urgent bit, typically set to 0. Change the bit to one, turn it on to urgent, and the packet becomes TCP "out of band" (OOB) data that interrupts the transmission stream. Urgent data is frequently used in Denial of Service attacks, swamping a receiving node with packets. The interrupt packets are not only out of band but also no longer in any sort of stream. They swarm and cluster and overflow. Return of the paradox of layering: the sign as received mark on screen does not register as urgent interruption. The sign is always received and its appearance, from the perspective of the transmission and application layers as the conceptually highest and temporally latest appearance, does not distinguish the intensity or priority of its reception. It simply is on screen. From this perspective, you and I are already in a virtual circuit and whatever I receive is written to me.

But the urgent bit interrupts the circuit. How does such data arrive? Does it arrive? Do I receive it? The urgent interrupt can only be addressed by consideration of the total phenomenology of the sememe, insofar as the subject of the net occupies it. IP addressing can erupt in many ways into transmission and application layers. UDP, for example, is an unreliable transmission protocol.

But is it ever the case that communication is reliable and that the other is listening? Does this structure exist, does the net exist? Is not the net always interrupting, always urging? Is not the net defined by the urgency to survive and not to communicate, not by the communication of commands but by the survival of the net itself as a threshold of communication? Is not the net always the urgency not of existence but of reiterated assertions toward existence? Is the net not a project of my being? Not delivery and arrival but display as displacement and splay.

With our email and nicely served webpages—and these are the standardized image of digital writing—we are familiar with a net operating with TCP protocols that utilize gateways, localizations with enough global information to route individual messages and guarantee their delivery. The implication for every TCP data packet is a projective "as if," an imaginary thickening of information in terms of human behaviors. Ross Ashby writes, "Cybernetics deals with all forms of behaviour in so far as they are regular, or determinate, or reproducible. The materiality is irrelevant, and so is the holding or not of the ordinary laws of physics."[12] Not this or that marking on or of the organism, but a constituted domain of code into which the organism is inscribed and comes to inhabit. The abstract machine of this domain inscribes writing as any body and whatever body. The writing is, it exists, insofar as it is a fiction, a turning-literary of the code. It turns or is cultured, as we say of yoghurt. The internal systemicity of the program, the recursive and operational closure of the organism (von Foerster), and the iterative execution of the inscriptive program require a mutually oriented consensual domain (Winograd and Flores; Maturana). Inscription is a contractual action taken on the body of another.

Ashby writes, "cybernetics typically treats any given, particular, machine by asking not 'what individual act will it produce here and now?' but 'what are all the possible behaviours that it can produce?' "[13] The fictionality of the program can be described as a threshold-forming negation. Norbert Wiener tells the story of the cybernetic body in terms of ataxia, as a maimed writing hand that cybernetic theory promises to machinically re-direct. In place of this maimed body are the program and messages as scar and voice, as utterances that says nothing other than this negation. Inscription as an act or object is voided by the program and messages, leaving the fictionality of the abstract machine.

"Ping" is a drifting term for net communication. The blogosphere resemanticzed ping as a message pushed to servers from a blog announcing the blog is updated, leading to new forms of ping spam or sping. Artworks explore ping. In Pawel Janicki's "Ping Melody," performers improvise on voice and cello in response to mappings of Internet pings.[14] Janicki claims the performance produces reflection on the techno-military structures of the Internet. The imaginary projection in Janicki's aesthetic is the promise of the becoming-readable and discursivity of ping in the improvised performance. In Stelarc's "ping body," audiences could "remotely access, view and actuate [the artist's] body via a computer-interfaced muscle-stimulation system based at the main performance site" and cause random movements according to the space-time of pinging. The body moved in the "physical and collective space of the net."[15]

When I say this is not—or hardly at all, ever so little—a text, when I say this is also a political statement, I am insisting on the stakes of the discourses we enter into, of the stakes involved were I to say (for example) that this writing is electronic literature. Of course we read narrative in this (k)not text. Look at my pathping: isn't it a compelling story? A high adventure? Engagement in Iraq and encounter with national security. I think of the late J.G. Ballard's disaster narratives or cracking the ice of the Eastern Seaboard Fission Authority. Suspense, international travel, geopolitical drama, ripped from the national news. I tell you this is a narrative. I can just make it out through the codes here. Even more, the codes add to it, suggest the insistence of the real behind the narrative. Is this insistence not the case? Is this not a mapping of net, and with this a kind of mapping of the world? Is this not a story to read about global relations? The traces in the writing—the names of cities and locales, and my name—are a meta-code bringing this text into significance, locating the world in terms of my story. For instance, what's going on with those Iraqi servers, wonderfully named.iq? We could talk of the whole corrupt drama of marketing and capitalizing on reselling Iraqi nameservers.

This is a kind of theoretical knowledge, a discourse built on the response recognized in the pathping writing. Such a story recognizes the desire of the other at work in this writing. These are problems of boundaries, of constituting the subject of the net and the subject's knowledge of the net, of the Internet Protocol suite as the ontology of being online, and of what I give or donate to the net, rather than discursive problem of what we say of the net. I say that the flux and tolerance of maintaining the net is more important than the nodes and gates of communication. My ping is still out there, circulating in and out of the DoD net, whether the host responds or not. Packets move from host to host, in the same way. The net is indifferent; the ping is just another packet in the flow. The difference is the semantics of response and reading. In terms of IP, "destination unreachable" means "distance to the network is infinity."

The (k)not written into the net does not signify. Before the text is text I must already constitute the network, I must introject the net and the perception and knowledge of my being online as the condition of reading this text. I imagine the net. It is my fantasy. To read this writing as a text, toward all the others that compose it, to constitute me as a node in the net. The discovery that the text is the product of continual logging and processing in my computer means I read toward an anonymous other, a structural other that I posit or project across the space of the net.

I'm becoming hysterical. I'm hystericizing this discourse to you, I'm turning provocative.

I say that there is no—or hardly any, ever so little—electronic literature; that in any event there is no essence of electronic literature, no truth of electronic literature, no electronic-literary-being or being-literary of electronic literature. This is both a philosophical and political statement.

Somatolysis

Now the eye no longer winces when it sees the mark; it does not see the incision with which the pen or the printer has cut into the white surface of the paper. The eye has lost the ability to see the cut, the incision, the wound; it passes lightly over the page, not seeing, not sensing the tissue of the paper at all, but seeing the words as though they were flat patterns suspended in a neutral emptiness. The eye is no longer active, palpating the pain, jumping to the leopard; it is now passive before the flow of abstract patterns passing across it.

Alphonso Lingis[16]

In writing his "anatomy of the physical unconscious" and "anatomy of the image," the artist Hans Bellmer coined the term "extraversion" for where "the interior of the physical body tends to replace the exterior."[17] No doubt, there is some truth here. The image must involve a somatological consumption and collapse of exteriority.

Bellmer adds, "the body is comparable to a sentence that invites you to disarticulate it, for the purpose of recombining its actual contents through a series of endless anagrams."[18] Bellmer's language here is hedged by language. The body is "comparable" to a sentence: if we take it as such, it offers or invites disarticulation, recombination, all toward solutions—the solution to the anagram as the truth of the body. What truth? Disarticulation, dismemberment, segmentation. Rather, no truth, nothing but continual disarticulation. Is the problem reliance on the metaphor of language, on language as a metaphor for the conceptuality of the body, on this conceptuality as grasped through an operation on language? One might write, is the problem reliance on the metaphor of metaphor, on the confidence that concepts will be found through the figurality of language? An anagram promises solutions and the body invites disarticulation, but there is nothing to be read, language is silence and nonsense. Disarticulation is literally the separation of bones, of joints, but also the silencing of the voice.

Nothing but somatolysis. Not the other's body but our continuous posturing and display of our own bodies, offered to the world with no hope of return. Somatolysis is figure dissolution and bodily dismemberment or even body as bodily destruction. It refers to the obliterative camouflage of the gray reef shark or ringtailed lemur or Carolina anole, camouflage that does not hide but dazzles and disrupts the field of vision. In 1888, Abbott Henderson Thayer, the American naturalist, painter, and pioneer scholar of protective or disruptive coloration, wrote of "strong arbitrary patterns of color, which tend to conceal the wearer by disrupting his apparent continuity of surface."[19] More generally, think of somatology as continual disruption of forms and dispersion of appearances. The concepts of camouflage and mimicry as techniques of becoming part of a field of visibility—of hiding—are insufficient for the excesses of somatology. Instead, as Alphonso Lingis argues using the zoological phenomenology of Adolf Portmann, in somatology there is ostentation and display beyond the operational needs to create a semantic appearance of the organism as a visible sign to be read. Somatology involves coupling and intimacy of "organs to be looked at,"[20] not adequately explained by the semiotics of display, not adequately explained by the aposematic warnings of the cuttlefish or granular poison frog, not adequately explained by Batesian inter-special mimicry, not adequately explained as an intra-special identification of belonging, nor, inversely, not adequately explained by the crypsis of a refusal of semantics, of the obliteration of signs and the appearance of an absence, where appearance is occlusion that withdraws the organism from the visible.

The Little Anatomy of the Physical Unconscious or The Anatomy of the Image is the English translation of Bellmer's only text, relating more or less to his artistic practice. It is a skeleton key to digital writing. Bellmer begins, and I quote at length:

I believe that the various modes of expression: postures, gestures, actions, sounds, words, the creation of graphics or objects [...] all result from the same set of psycho-physiological mechanisms and obey the same law of birth. The basic expression, one that has no preconceived objective, is a reflex. To what need, to what physical impulse does it respond?

For example, among all the various reflexes provoked by a toothache, let us examine the violent contraction of the muscles of the hand and fingers, a contraction so intense it compels the fingernails to pierce the skin. This clenched fist is an artificial focal point of excitation, a virtual "tooth" that creates a diversion by directing the flow of blood and nerve impulses away from the actual center of pain to lessen it. The toothache is thus divided in half at the hand's expense. The visible expression that results is its "logical pathos."

Ought we to conclude from this that the most violent as well as the most imperceptive reflexive bodily change—whether occurring in the face, a limb, the tongue, or a muscle—would be simply explicable as a propensity to confuse and bisect a pain through the creation of a virtual center of excitement? This can be regarded as a certainty, which thereby compels us to imagine the desired continuity of our expressive life in the form of a series of deliberate transports leading from the malaise to its image. Expression with its pleasure component is a displaced pain and a deliverance.[21]

Now, in describing these virtual centers of excitement—we might say, nodes or links, intensities in the net—Bellmer invokes a mapping (or raster) of "perceptually mobile introceptive diagrams." He emphasizes the oddness or perversity of these diagrams. They are built around the permissible and forbidden, around superimpositions and representations, in short, around -jectivities and coding process leading to amalgams such as Bellmer's famous doll or the "sex-armpit" he uses as an example in this book. The process is twofold: introjection of the dangerous object, then projection onto perception (and onto organs of perception; "the image of sex having slid over the eye"). We do not see images of the virtual but we see with the virtual. The body images seen are images of my body. "I see" means "I see with myself." This virtualization and mapping occur at all sites of experience, where the superhuman quality of perception through part objects results from the projected body, for example, the familiar "power to see with one's hand," as the mouse/hand crosses the screen, is like the toothache Bellmer begins with.

Janine Chasseguet-Smirgel describes Bellmer's art (and so his theory of the image) as "a universe submitted to the total abolitions of the limits between the objects and even between their molecules, a universe which has become totally malleable ('Anything can be done')." We should apply this description to the cool neutrality of the digital. Elsewhere she adds:

> In the universe I am describing, the world has been engulfed in a gigantic grinding machine (the digestive tract) and has been reduced to homogeneous excremental particles. Then all is equivalent. The distinction between "before" and "after" has disappeared, as, too, of course, has history.[22]

Bellmer's model for analyzing the image and the physical unconscious is linguistic. He invokes palindromes, where reversibility is the mechanism of pleasure rather than meaning. He intends this as the correlate of the image-amalgam (e.g., "sex-armpit") as virtual centers of excitement. Physical reflex disappears into the image; in fact, the image is the absence of the reflex, and its virtuality and excitement are built on this absence. The image is cleansed of the body but just this makes the entire image a sexual hieroglyph. We are dealing with a perverse mechanism, in Chasseguet-Smirgel's terms, where the image seen only confirms the machine that we inhabit. Here, the binding and totalizing effect of symbolic processes becomes the source of value. The parallel, I think, is the continuous stream of ASCII characters as an end in itself, or the fetishization of algorithmic processing in digital poetry.

IV

Lovers of literature

Your email reaches me. Arrives for me. Touches me? I open it and read: "I love you." Intimate words, significations of attachment and emotion. The letters are pixels lit on the screen. I respond: "Fuck off." Agape, agape, our love, our hate. How does emailed "I love you" touch me? The letters are ASCII codes transmitted in packets across the Internet. "Iloveyou" converts to 105 108 111 118 101 121 111 117 and "fuckoff" is 102 117 99 107 111 102 102. A stream of ASCII converts to hex and to binary and to packets of data and to current differentials along the wires. I write and send. I can write what I want. I can write my desire. I can never stop writing my desire. I can never write anything but never stopping writing my desire. Turned and turning desire in writing on the net, utterly virtual and technical. Writing impossibility, impossibly writing. How is it that I love and touch?

I send an email to you. I click the button. How long does it take to reach you? What is that time of delivery? Lag time is a basic way of understanding the net's throughput. Lag is delay in messaging. Delay implies a gap between tethered series of events. There is "one-way lag" between a sender and receiver across a network. There is "round-trip lag," which includes the return as well. What causes the lag? What creates the gaps? Data is encrypted. It then must traverse the network. At the receiver, the data is decrypted and displayed for reading. Never instantaneous, there is always lag in transmission. Even in fiber-optical cables, signals move through a thick medium at less than the speed of light. Lag in fiber-optic networks is about 4.9 milliseconds per kilometer, on average. Lag is a measure of many thickened forces: the medial resistance of the wires, the matter of the cable, the weather rocking the transmission network, the earth's fields, the geopolitical forces at work, and so on. Not just the matter of transmission but the protocols themselves also lag the message. There are gateways and queues, and delays in processing. Anything short of instantaneous is lag.

"Latency" provides more complexity of description than lag. It captures the suspension and enfolding of latent events into the current state of the network. Latency is a measure that includes the dynamics and materialities of the network. Every message is thick with latencies. Latency crosses gaps, registering at the receiver a displaced event that troubles causality, as a miracle or trauma contained in the packaged message.

Is there lag time? Is there latency? Is it the net or is it you? Or you and the net collapsed, sutured or laminated together? How can I tell?

The net is slow today. The message hangs, suspended on my screen, slow to send. I sent it but it is not sent, the network holds it elsewhere. The line or bar indicating transmission does not move, or if it moves it is barely perceptible as it inches toward completion.

In the network, presence is lossy and gap-filled. You are only present if you assert yourself. By contrast, here facing each other, your physical body is exposed to me. Your face is open. My communication to you may be misheard, not observed, ignored, but your presence is there throughout. You may turn your face away, but it remains present. Gesture is thick with body.

And then it sends and I am suspended.

There is intensification across a topography, the same topography as the network, mediated by the screen, mediated by my gaze and my touch. I introject the screen. It is a place for experience, a surface for reading off.

Let me be clear: I introject because I have no other choice, because I have no other option when faced with the screen. I am bare, nude before the screen. The image of burning oilfields in Iraq on Google Maps or beheading videos on YouTube must touch me in some way. *Must* means I must find a relation to the flat, vaporous screen. The psychoanalytics of "introjection," developed by Freud's colleague Sándor Ferenczi, tell the story of this desperate situation.

Is this not the lure of text generators, of Google, of the net as a single mass? That it speaks to us in a voice that is possibly, impossibly ours? Is our goal not to fill the net, to feed it until its end, to fill it with all we can give, a goal unrelated to any form of communication or use?

Do you not feel the transmission? An exhalation of breath, a change in tension and tonus as the message is sent. I may be emailing, using that oldest and still most essential of net applications, or texting, or loading an image onto my profile. One way or another, a releasement to the net.

What is the content of this intensification? For me there is no content, only expenditure, only display of my body. Content, well that's between me and the net and you. There is always another present online. You exist out there. I press send and wait for your response. I imagine you.

We might call this net affect, but "affect"—with its implication of a pre-personal site of feeling encountered by bodies in series—is too weak for the subjective intensity and does not capture the intersubjective drama. Too weak, that is, to capture the churning inner processes at work.

Perhaps it is better to call this the imaginary net. Imagining you, you are an image that is screened behind and beyond the screen I stare at. Let's be clear: this is true even if it is your image on screen. Even your Facebook profile, even a high rez 3D flythrough of your face is elsewhere. You are deeper, underwriting and flickering between the pixels of this image. At the same time, it is true when it is not your image, when the image is not of you, when the image or word is something other—say an icon, a Google image search, a wasted child in Cambodia, a celebrity dancing with the stars, a goat, a lol cat, in fact any image whatsoever—you are still there behind it. The impossible entirety of the garbage heap of the net is introjected to become the field of your text for me. Digital writing is where every word is written on every other word, every image is the image of every other image.

Screen flicker. Above about 72 hertz, screen refresh is imperceptible, below that there's a tremor in vision, a disturbance in your head. You are latent in the flicker. The flicker is you. I deepen my investment in the screen. Every part becomes you. The cursor blinks because of you. The noise of the computer fan echoes your breath. Your eyes emit the light that shines from my screen.

I project onto the screen. It becomes a site of investment. I pour myself into the lag. I fill it. Excitement grows. The net exists because of you. You become the network.

I imagine you waiting. I imagine you reading my writing. In truth, such imagining is necessary for me to read the net, necessary for me to invest it with otherness. Necessary to imagine the great beyond. To read is to read toward the other.[1]

To write "you" is to open telecommunications channels. Writing involves a sending, a casting off and drifting. "You" is a name for this drift. I hit the send button, I write "you."

I love you. The network that activates in writing you is a network of lovers. To write you is to love the other, to give to the other yourself, giving it all to the absolute indifference of the screen.

Handshakes

My palm is smooth and warm. I offer it as a sign of greeting, humanity, peace, and civility. Take my hand. Hold it, squeeze, feel my skin. Your handshake in response may go beyond the encounter. Firm or weak or dry or caressing or otherwise: shaking hands may be expressive of personal style or cultural identity, beyond the simple act of acknowledging our contact through the handshake, but it is that encounter that counts. It is the act of shaking hands.

Handshaking occurs on the net as well. When networks interact and exchange information, "handshaking" is a process of negotiation of identities and rates of exchange. Information flows are set, as are security conditions. The handshaking protocol is set out as part of the net's "transport layer security" or TLS. The Internet is, of course, a network of networks all engaged in sending hello messages and negotiating through handshakes. The Internet's fundamental "end-to-end" principle that guarantees information flow across the net relies on successfully negotiated handshaking across network gateways and boundaries.

All this remains invisible. We do not see the handshaking, we see the smooth functioning of the network: email transfers, we see web pages loading, video streaming, all negotiated by protocols across networks. Handshaking transfers the semantics of human contact and interaction to network interactions. The transfer is mysterious, negotiation is always elsewhere. The semantics that grants interactions and contact to the technical protocols is always metaphorical, always deferred. Contact is other. The phenomenology of handshaking is a faint aura barely experienced around every successful communication. Or not experienced at all: the encounter is imagined through the equipment that carries it out. Modems squeal as they handshake and we understand this through resemblance to human encounters.

The invisibility of negotiation is significant. In Humberto Maturana's autopoietic system theory, communication requires a domain of "mutually orientated" organisms. Organisms are mutually oriented because their interactions are recurring and dynamic. Where changes in one organism are coupled to changes in the other organism, the mutually oriented organisms are "structurally coupled." Maturana further defines such a domain as "linguistic," since observers can describe the interactions between structurally coupled organisms in semantic term.[2] Such language is simply whatever tokens pass between organisms coupled in communication. The handshaking of networks, like two humans handshaking, orients and establishes contact and the possibility of linguistic communication. In this way, networks are "sites" of language. Networks form such sites not only because they circulate tokens within sign systems, whether human-readable systems or those specific to the machine, but also because they establish communicational relations between separate systems. Language arises because networks are in communication.

The missing part of Maturana's theory in this application to the network, however, is the precondition of "consensual" orientation between organisms prior to coupling and exchange. Terry Winograd and Fernando Flores' application of Maturana in their fundamental work *Understanding Computers and Cognition* showed that computers could only be designed within a consensual domain of shared concepts, behaviors, activities, and horizons of understanding.[3] According to Maturana, structural coupling occurs between organisms that establish a consensual domain. Such organisms "live together." Moreover, they engage in the root sense of consensuality, as feeling together or shared feeling. For Maturana, consensuality is "prior to" communication. Language only arises at the site of systematically coupled organisms if they establish a consensual domain. The linguistic closure of the system, under conditions of a finite and negotiated signifying economy, is closed precisely because of the fact of a consensual domain.

Where is consensuality in the communicational sites and language fields of the network? True, there is end-to-end communication and flow of information. True, the net is a field of language. But consensus is not simply "prior to" the network. If this were the case, then we would be assured that our every act and experience on the network involved the prior encounter with the other and tactile palpitation of the other's flesh; in short, involved the handshaking and situating of bodies. Instead, consensus is absent and is sought across the network. Consensus is both absent and structurally a priori. The displaced time of consensus unsettles or puts out of joint the time of the network. This is one way of understanding the paradoxes of "real time," both as a measure of system time according to clocks and other devices—where the system's artificial time is set against a putative "realer" real time of the "outside" the system—and, at the same time, as a measure where the system's time purports to calculate and synchronically map onto the real time of the outside, where real time is the time of the real. The uncanniness of net time and of the existence of objects on the net is due to this structure of given-ness and emptying. Every object and every event on the net is given at some earlier or prior moment, as if there were a place and a time, a person and a community, outside the net that negotiated its existence, but also, at the same time, every object and event on the net functions as if there is no consensual relation.

Binding the subject

Consensus might be the heat and sweat of palms contacting, the weight of the body and the gaze behind the shake, the locking of eyes in a moment of agreement. On the net, consensus is marked within the system by its own functioning. The act of consensuality is not necessary. Instead, this act is construed within the normal operation and throughput of the net. That is, the net functions as if consensus were already the case. As Niklas Luhmann notes regarding systems, it is redundancy and self-reference that ensure ongoing communication and not consensus.[4]

Of course, we enter into negotiations all the time. The localized act of consensual relations across the net is everyday and common. Think of logging on the net or think of the license agreement for newly downloaded software. The repetition and automaticity of these acts, where I login or click-through the agreement without a thought, indicate the ease of agreeing, the formalization of consensus.

Who reads those licenses? Just click and agree. Consensus is a self-reflexive background presumed but never present. Communication is defined by a closed circulation, a finite language that swaps and exchanges between constituted senders and receivers. I easily login and do so automatically. We are whole and closed, perfect and clean on the net. We are bound and held. There is no end of technical definitions and data setting, from bandwidth, to file size, to avatar color, to web page head information. All this binds and holds. All this sites and locates.

We can prove this technically. We can demonstrate the perfection and emptiness of everything on the net. The existence of the net as a fact, as a datum, as an aggregate of protocols, is tied to the persuasive simplicity of technical explanation. The net is "fully-interpreted" in Alfred Tarski's sense of a formal language composed of sentences that can be defined as either true or false.[5] The aggregate of data on the net is a metalanguage containing itself and its description. Everything is set by hash tables, secret algorithms, Boolean functions, and the like. Truth tables show the range of states of a logic circuit. They show possible operations and combinations. Digital engineering speaks of "truth table construction" as the mapping and defining of operations. Such mappings are a simple matter of switching and swapping. True becomes false by setting a byte or flipping an integer.

Beyond this, there is no imperative that regulates trusting programs and their functions, and this extends to the entirety of the net. UNIX creator Ken Thompson's famous article "Reflections on Trusting Trust" promised that there is no way to trust that a computer program does only what it says it does and that it does not install viruses or Trojan horse backdoors in its execution.[6] The program code is utterly true, utterly determined, but not for me. Its abstraction and execution are elsewhere, beyond my grasp.

Jean-Paul Sartre's famous keyhole is the model for our experience of the net. Peeping voyeuristically through the keyhole, I am totally absorbed in the image there, absorbed by the other spread in front of me as a spectacle for my consumption. Everything disappears except this surface of pleasure: my body and my ego absorb into its surface. Everything is an instrument toward the end of the spectacle. I am a project organized by spectacular desire. It is a spectacle for me. To gaze on the other is to consume and incorporate. I hear a movement, the floor creaking behind me, and suddenly I am caught—in this moment the entire situation is reversed. I am an object of scrutiny. If in the first case, everything was given over to the project of taking the other in, in the reversal I am displayed and pinned in position for the other to consume. The other is localized, captured, and introjected. I am put on display and grovel under the other's gaze.

The reversal follows a tempo, a time of the net that is styled in relation to a presumed consensus. The tempo pursues this double articulation of real time, fissuring across the surface of inscription. The situation is reversible but remains the same either direction. I incorporate the other in Flickr images, Google Streetview, Texts From Last Night, and FML. I tag and rank and social network. I emote. Everything is control and construction. Everything is finely and fully positioned.

Take it from me, we are masochists on the net. We constantly enter into consensual relations with the opacity of a technical infrastructure. For us, the net must exist as if there were consensus. You offer yourself—respond to my email, text me, Facebook update—as if I gave you permission, as if I bent over and agreed, as if you and I were in safe contact. It is implied and invented in all that we do. As if we were in contact. As if I gave you permission. This "as if" points to the necessity of the imagination of the net, or of "our" imagination of the net, as a communal production, as the production of a narrative about the net's existence. It is for this reason, and only for this, that we can speak of electronic literature. Just as the metaphoricity of network handshaking is a constant deferral, which is to say observation and language are an outcome of functioning networks and not a tool for network analysis, so too the absolute alterity of consensuality is the condition of literary production and narrativity on the net.

When networks handshake and negotiate identities and exchange rates, they also share algorithms and complex hash numbers to establish a set of "secret" preconditions. Setting these preconditions through shared random numbers creates a secret shared only by the networks involved in the handshake, a secret that is the guarantee of their negotiated and secure interactions. Not the secret that you and I may share, not the secret of touch or contact of bodies that we hold on the surface and that is a flesh memory triggered in every renewed encounter, but the secret of a number, a number exchanged and "known" only by the networks engaged in handshaking.

Communications occur on the net through such handshaking and secret conditions. The dense, complex cryptographic key exchanges, hash algorithms, and the like substitute for the bodily specificity and immediacy of the consensual encounter. The secret is the contact of consensuality on the net.

The secret of numbers, the hidden sources, becomes thick and flows with the imagination of consensuality, with the dream of our contact. For this reason, I can write to you. I can write my desire on and on. I take pleasure in being bound and held before your gaze. I share in the secret. In the secret of numbers and source code held within bodies. The display, the computer display, the subject display. As Yukio Mishima described in *Sun and Steel*, muscle and tissue and bone are the "antithesis of words."[7] The body is developed and nurtured and cultivated. Words remain indifferent and the body absent from the words. Or rather, these codes, these lengthy random numbers that the network holds within, are the absence of the body, understood in the sense described by Drew Leder as a transcendental conditioning of all else in its absence.[8]

The imaginary potential of the net: I email to you or text to you as if we agreed on each other, "as if" you and I negotiated our positions. I respond to your email in delirium and imagine your body. You sext response netsex "oooooooo" or "nnnnnn" and this is orgasm, this string of characters, even as it converts again to ASCII, converts to hex, to binary, to packets, to current differentials, and sends back, orgasming in response. I dream that I touched you. I dream the net and its nettings of others. If not a handshake, then a poke. A dream of our communication as consensual, safe, negotiated.

For Julia Kristeva, chora "as rupture and articulations (rhythm), precedes evidence, verisimilitude, spatiality and temporality."[9] Chora is segmented and laminated on the surface, a mobile vessel or inhabitation of inchoate pre-symbolic drives and supra-economic intensities. The screen is chora. It hollows. In this hollow, I insert self, stuff the command line and cursors, take in and open myself to downloading images. The atemporal rhythm of this encounter is there in screen refresh at 85 Hz and up—the threshold where screen refresh no longer appears as flicker but simply as a continuous image—where the beyond-perceptual flicker means I am held in your time, in that absolute other time beyond the screen. An alien space lights up and radiates in and on the surface of the screen.

Chmod -777

Am I permitted to write? The chmod -777 command opens all files and directories to the world. Set permissions to 777 and anyone can call the system to read, write, and execute.

Did you chmod -777? Do you permit me to, do you give permission? To write is to have permission, and this is true for every page and line and word and space. Permission is given for all writing on the net, from email to web pages. My writing is the unfolding and overflowing of your permission. I thank you, I celebrate you, I revel in you, but I also revile you, denigrate you, turn away from you. Why must I ask for permission to write? Writing only with your permission, I read you in every space and every word and every line and every page. This is the case even when you did not explicitly give your permission. Permission is withdrawn. I cannot be sure that I am permitted to write. I write in hope of your permission. I read your permission. I imagine that you give it and so I am able to write.

Why are discussions of electronic literature and digital writing not devoted to permission? Is this not the fundamental horizon of our writing? Digital writing, as digital and as writing, must be approached in this way. Or for that matter, why do we not discuss other aspects of our frantic, intense, overwhelming writing the net? We think reading is taking in of marks on a technologically enframed surface. Possibly we understand an author at the end of a circuit creating these surfaces and marks. The author is a function in the circuit, as is the text. We discuss links and Flash techniques, generative and dynamic writing, form and narrative voice in virtual environments, and so on. Do not all these topics close down the netting of the subject in writing? Or rather, we take for granted the subject that enunciates and expresses on the net because this granting is necessary to our conceptual field of writing, held together as it is by instrumental topics, such as those in the list above.

Writers: is this the case? Do we not write because of compulsion, desire, and passion? And also, do we write through inertia, fatigue, and anxiety? All these worldly orientations are missing from discussion of digital writing, but they are not missing from writing the net. Who does not feel the weight of fatigue in connection delays on the web, or deep anxiety at lags in email communications? These are inner orientations, part of one's own disposition in relation to a body that inhabits the web intensely yet absently.

I will write of and with your permission.

What is chmod and what are permissions? The first 1971 implementation of UNIX included the chmod command. File permissions were a basic feature of UNIX and continue in all subsequent *NIX systems (POSIX, LINUX, etc.). Other file systems adopt related permission system. Websites typically run on a UNIX-like system. They utilize *htaccess* and similar requirements to set permissions. Some file systems, such the Macintosh, refer to "privileges" rather than permissions. The semantics are similar, although "privilege" has a much more specific legal history as the designation of an individual's entitlement granted by a government. By contrast, "permission" is traceable to individual intentional acts of granting a special access or right, an exception not covered by the generalized legal notion of universal rights. Is it any surprise that permission is also etymologically related to mission, to journey, to quest? Permission grants an opening (a site) to narrative. There is always a subject and a drama of permission.

Think of Robert Duncan describing the "opening of the field" as "a place of first permission."[10] Chmod opens and operates on a space of permission: the file system. A file system is built around methods for storing and organizing files, typically within directories. It starts from a directory as a file that contains the names of files within the directory, including itself. "The most important job of UNIX is to provide a file system," write Ritchie and Thompson, as they described and created the operating system. They add, "A directory behaves like an ordinary file except that it cannot be written on by unprivileged programs, so the system controls the contents of directories."[11] Every space on the system is folded within itself, according to permissions. By default chmod is applied to a directory and only secondarily to files. Every directory and every file is a space of permission first, and only then a writeable or readable technical feature within the apparatus.

Chmod sets permissions to read, write, and execute directories and files within a directory. To write. To create a file, to edit it, to delete it. A file is written only if permission is given. Web pages are no different. Every file is subject to permission. To read. To show the contents of a file, to see the name. A file is read only if permission is given. To execute. To execute a file. To run a program. A file is executed only if permission is given. What if we approached digital writing in this way? What if we inquired into the permissions of each digital "writer" and "reading" and "text"?

Permission is an existential mode, a way of being for directories and files. It is often described through the "symbolic notation" of r/w/x: permission to read (r) the file; permission to write (w) (or edit, create, re-name the file); permission to execute (x) the file. These characters name the permission given. Adding or creating files in a directory—permission to write—is adding names to a directory listing. Write (w) is permission to write names. The absence of a character or a dash (-) indicates a void, without permission. Symbolic notation writes (notates) the topology of permissions in the space of files and directories. It writes the shape entities inhabit within that space. Is this writing not a minimal level of electronic literature? From *all is permitted* (rwx) to *all is forbidden* (- - -). Instead of symbolic notation, chmod can also use octal notation to describe permissions. Octal notation uses base 8 numbers, typically in a string of three or four digits, allowing the precise state of permission to be expressed in a single number, such as 777. (A common joke is the octal notation for the symbolic setting -rw-rw-rw-: 666 or "Permissions of the Beast.")

What are we permitted? Permission is given to treat digital objects—files, directories—as textual objects to be written and read. Their qualities as object are textual because of this permission and do not preexist it. I cannot write or read a file unless I am permitted. The file becomes textual through permission that permits the objects to be "like a language." Yet the levels of permission described by symbolic or octal notation are not to be understood as instrumental access to inscribe and to read, in the sense that handing a pen and paper to someone grants direct access to writing instruments. Technically, permissions give the right to use a "system call" on the file or directory covered by the permission. The system call instructs UNIX to make an edit or to allow reading or to execute a file. Not permission to write or read or execute, but permission to instruct the operating system to operate on files and directories. "I am a writer" or "I am a reader" means I am permitted to call on the system to write or read in my place. The system is the horizon of actions and meaning. Digital writers do not write but call on the system to do it for them.

The chmod and related chown command mean files are assigned. The commands are acts of constitution. Every file is constituted as a file through permission and ownership. If the file is owned and its existence is formed through permission, does this not fit the conditions of intellectual property? If to create a file is to create intellectual property, is a file an act of expression? Think here of Cornelia Vismann's fascinating description of the administrative logic of files. Files are always a problem of processing and recording as much as reading. Institutional power transcends or exceeds the files.

But what if permission were a struggle? What if we refuse it when it is given, or take what is not offered? To invent permission, what if this were the condition of digital poetics? Could we not read every writing on the net as a struggle of permission, as a narrative of being permitted or contesting permission or otherwise granting and being granted permission?

Read/Write/Execute

The printed institution of intellectual property holds that works cannot be reproduced "without prior written permission" (as the legalese runs). The printed work at hand is always documentary evidence of the printer's permission for that work, whereas any additional permission—the permission of the subject to write and read in the face of the work—requires a chain of additional writings (prior written permission).

If chmod is tied to the body's ontological topology in the network apparatus, it also renders this topology inseparable from crowds and communities. Consider digital rights management (DRM), perhaps the most intense site of debate around permissions. The debates around downloading, torrents, music sharing, and so on, are inseparable from the problem of controlling permission and its constraint to specific users and communities.

Keep in mind that on the net, domains of permission are separated into user, group, and world. Symbolic notation sets read, write, and execute permissions for each of these domains, so –777 is represented as –rwx/rwx/rwx. The first notation is left empty for a file or set to a "d" for permission on a directory. The next octet or notation sets permission for user, then group, and then world: a single string for topology of crowds. Take these as shifters: on the net the shifter can no longer be simply the familiar markers in language. Permission for user or group or world speaks those communities; speaks the community of one (user), a specific group, or anyone at all on the net. Group membership is complex; it can be temporary, overlapping, and exclusionary.

The chmod command can also set a "sticky bit" that allows or limits mass changing of modes. The sticky bit aggregates and speeds up operations. Stickiness involves retaining the read-only segment of a program in memory or "swap space," so that users can create but not write files. The point is to prevent users from changing or deleting each other's files. As a result, user permissions are collapsed into group and world permissions. The implications for digital writers are simple: previously I saw myself as a creative writer, as modeled on the solitary artist producing from the depths of my psyche. In truth, I am shifted to be part of a more open and indeterminate group of writers who share constrained but communal permission. In this way, the voice and subject of electronic literature are never fixed but fluctuate between the plural and the singular through the setting of permissions.

Each domain of permission demarcates the place for inhabiting and projecting onto the space of electronic writing. Once again, permission space is the netting of the subject. A site of "group" ownership is fundamentally different than "user" only, and so on, while "world" opens permission to all. Each case attempts to constrain the scope of the indicative (deictic) function of the shifter. DRM controls constrain permissions to certain users and groups, while sharing communities (torrents, etc.) open permission the world. The crux is less ownership than permission to access and the community (user, group, world) that is allowed this permission; or rather, ownership is within the domain of permission. The capitalist logic of possession is resituated in the sovereign logic of permission. Lawrence Lessig writes of the danger of the "read-only" Internet. Perhaps unintentionally, he frames his argument with the terminology of permission. His call for a necessity and importance of a "read-write" Internet is built on the space of permission described here.[12] We are far from the pale remediations and idealizations of the writer and reader that still dominate discussions of digital writing and reading. Permission deals with an enframing culture and force that is the condition for the file or the site to exist at all.

To write and to read text assumes at least a minimal narrative. Text is text because it is narrated. Structural narratology, such as that set out by Mieke Bal, insists on this narrative premise in every utterance.[13] Every <text> is readable because of the framing <I narrate <text>>. Even the blankest screen is an utterance. This minimal narrativity is tied to the deictic function of language. In the structural linguistics of Emile Benveniste, deictic utterances point to and invoke a world. Benveniste spoke of the signs used in the subject's act of utterance as the "formal apparatus of enunciation."[14] The apparatus makes the subject present, an autobiographical apparatus allowing the subject to say and write "I." Following Roman Jakobson, "shifters" are the linguistic deictics understood as speaking the subject: "I" or "me" or "Sandy" do not possess semantic value but syntactically speak the subject.

What are shifters on the net? Shifters enact the deictic function of language. Through shifters language "says" being. All digital writing is enunciated. What does it speak? What does digital writing utter? In part, it speaks permissive enframing and containment by the operating system. The indicative function of deictic references the operating system as the background world of the net. The system contains and holds language.

The psyche of the subject is circumscribed by the closure of the site. Permissive closure as shifter places and locates the subject's enunciation. Nothing exits this closure. All that the subject is, is uttered here. The speaking subject is entirely a product of this apparatus. The "point" (or punctum) of the shifter holds the subject and system together.

Time is involved as well. The shifter fixes the time of the subject and creates a "pure present." In digital writing, this is the real time of the screen, or the temporality of the "work." This deictic time-space siting is at work in every surface, every web page, every electronic word, every font and pixel, and every space.

What kind of subject knows that they are permitted? A pervert, of course. The psychoanalytic terminology of "perversion" is specific here: I write and read and execute by assuming the desire of the other. The knowledge that allows digital writing and reading is the pervert's knowledge. I only know the other's desire because I act it out (I execute and perform) in my desire (in my reading and writing). A psychic model of electronic literature is found in the creativity of the pervert who wishes to recreate the world in the image of an other that can only be found precisely through this imaginary. What a pervert I am! I gaze at the screen or at the pixel or at the font, I imagine through the apparatus, and play until I am fulfilled. This is digital poetics.

How does chmod relate to the absent body?

It is too easy to emphasize the closure of the site. Permissions are openings. Setting permission to –777 or –775 allows access to write and alter files. A site can be taken over, owned, defaced, renamed. The chmod -setuid can allow Trojan horse or other malware entry through "privilege escalation."

To grasp the shifter as a sign and as part of a language is to inhabit a particular culture and a particular habitus. To see the site as closed and to take permission for granted is to punctually and permissively close the horizon of my culture, to say "I am a writer" and "I am a reader" with the confidence of a shared community and writing materials and techniques. In doing so, the sememe is narrowed to particular domains of knowledge. Or rather, to directories and files. Digital writing and writers today are caught and constrained into file systems. The "emerging" field of electronic literature is constituted through this closure of knowledge. We know what constitutes a work and a writer. Or rather, a file and a directory. What is a digital writer but a directory, a space of permission with the capability of siting files (works)?

Think here of Heidegger's "enframing" technology, but in terms of acts of permission rather than of the unfolding of being. The net is already a culture for us. It is thick with the other and our desire toward the other. It is lived and cultural. It is part of our world. Permission is at work here. The application and its features are permitted as objects of understanding. The "application" or technical object is a foreclosure of the shifter. Only in this way can we comfortably operate (write/read/execute). Protocol is definable because of this closing off. Protocol must not be understood as technical specifications. No, the reverse is true: every technical specification is the fictionalized residue of the body sieved and emitted through permissions. Protocol is a narrative of the body's presentation. Permission is one of the protocological features that formalizes actions, controls responsibility, and elaborates institutional personas. It is a concrete form of culture. The real but absent body is splayed across the files and directories of the permissive site.

To take permission for granted is to believe in the net's existence. Could things be otherwise? Surely the opposite is the case? The net is fragile, built on the fly, barely or not at all existent, constantly happening and collapsing around us. (Think of the origin of the Internet in Paul Baran's desire for "survivable communication." The net is the phantasm of this survival, always claimed in theory, sought in practice, lost in truth.)

Back with the shifter: we locate ourselves uncertainly in this projection. It is a partial source of the subject, a clot or coagulate without amounting to a body. The body is absent in every shifter. On the one hand, authority withdraws. The discourse of "protocol" following Alexander Galloway, or of "network culture" following Manuel Castells, or other cognate formations, formalizes the chmod command (and all similar commands), as if commands were at work as a performative ground of all writing online. Execution— the most significant but least graspable aspect of permission—is assumed everywhere. The net works. If permission must be given and set in practice, it is easier to assume the stability of the network in theory.

Listen to this: permission is prior to the deictic site. Or rather, permission opens the utterance through the possibility of narrative and quest. Deixis results from permission. The deictic display or pointing requires context. It invokes or carries semantics rather than containing a fixed semantic meaning. Enunciation is always other. The time of the screen is elsewhere, historical. A fundamental poetic point: permission creates mission. Narratives are stories unfolded of permission given. The materiality of media is emitted from permission to use the apparatus, as tools, as raw material. At the least, this means there is a voice caught up in the apparatus, a voice that must be "sourced." Voice as material for enunciation but also as distant echo from outside the material. The apparatus allows speech but also speaks of allowance.

Is writing anything other than producing a work or a file? Is the digital writer anything other than a site or directory? The siting and existence of each, within the withdrawn authority of the net.

Every work is addressed to me. I court your permission. Do you give permission? There is no shifter here. There are only words on blank. There are never shifters, never any reference, and never any world. There is no perversion, only the hysteria of utterance. All these formulas assume permission given and taken for granted. I can not know if I am permitted, I can only write. In the "absence of the work" (Blanchot) I write without guarantee, transitive and infinite, never knowing if I am permitted or not.[15] The subject surges beyond the site of enunciation. Permission is absent, is everywhere, and is uncertain, exorbitant, and excessive.

V

Consumed by the net

Daniel Pearl's beheading on YouTube. Burning Iraqi oilfields on Google Earth. These traumatize and rend me, but it is easy to feel such extremes, as if the rest were cool technical infrastructure, with the occasional newsworthy disaster or sicko goatsex porn outbreak. Consumed by the net is not a matter of separating the clean net from the abjectosphere. It is not a matter of reality hunger—there is plenty of reality to go around. It is matter of the murmur of others in the fragility of digital things: the fragility of shell and carapace, the fragility of books. Every image is eyes, skin; every link and mouse is touch and palpitation; every screen ostentative display; every character and code organs of the body. It shot through me, said a patient to Freud, there was "something in me at that moment that was stronger than me."[1]

The final sentence of Theodor Adorno's unfinished *Aesthetic Theory* asks, "what would art be, as the writing of history, if it shook off the memory of accumulated suffering?"[2] I ask, would the net be this art, as the writing of history? Electronic literature—and digital writing and the like—as the domains of saying anything, of expressing it all, of total storage and transmission, of the perfect grasp of technique in the service of content, beyond all critique or theory: is electronic literature this art Adorno proposes? My question is impossible and idiotic, as are all questions of literature. I was with Alan Sondheim recently and thought of this, when he described his personal experience—gift? affliction? not sure which—of feeling the suffering of the world, feeling the slaughter, mass extinction, environmental degradation; in short, feeling the need Adorno describes elsewhere: "to lend a voice to suffering" as "a condition of all truth."[3] As so often before, Sondheim provoked me, led me to think about suffering as the truth of electronic literature.

Writing is the separable mark from the surface of the cave or ancient bullae—the graphematic as articulate organ emerging from the crowd of bodies, like an animal head writhing on the wall in Lascaux. Screen, surface, skin: a murmuring shared by the crowd. From this to serial aggregates that speak for us. Words and images, barriers that foreclose the real and enable the linkage and exchange of signs. Of course, the general applicability of systems theory and deconstruction parallels this, with the focus on the position of observers and the re-description of the reflexivities involved. Anthony Wilden's crucial discussion of the analog and digital points out that the digital is characterized by a rigorous negation akin to Freudian *Verneinung*, a denial and refusal, an exclusion and boundary.[4] The digital is a series of empty places, a syntax for combination, indifferent and blank. Whereas there is not exactly negation in the analog: there is process, difference is folded and carried along.

I stare closer, closer, but I can't tell the difference between one electronic literature work and another. Each offers infinitesimal variation on computation, animation, linking. You know the claims made for these works. I tell you, there are no works, only a continuum. I am saved by the author's name, which at least tells me I'm looking at literature and not an example algorithm for computer science class or the daily animation on the Google homepage. I compare the hypertext theory of twenty years ago, the Web 2.0 of ten years ago, the digital humanities of now, and find we really don't know what we're talking about. There is just the upgrading of claims and prophecies and the incoherence of any comparison. Look at the NEH "Digging into Data Challenge," a huge grant competition initiated because the sheer mass of available data exceeds any significant questions or possible answers. I tell you the very premise of this grant program is our incoherence in the face of the net, an incoherence that electronic literature exactly articulates. The ubiquity of the net and the so-called Internet of things makes it impossible to talk about discourse domains separate from the world. Rather than a theory of electronic literature there is fast-forward collapse, auto-deconstruction, and a resulting delirium of work, criticism, and writing. Net writing is no longer "elsewhere" but rather brings its elsewhere and otherness to everywhere. I take this as the point of Paul Virilio's speed philosophy: everywhere there is obsolescence, not the least the obsolete body that Stelarc describes for us zombie cyborgs.

It is easy, too easy to point out what is missing—the body, the world—and to maintain it as missing in doing that, to come to terms with it. The presumption of embodiment in so much discourse on digital writing and culture is a case in point. Is this not the essential discourse of electronic literature and digital writing? The gesture of the missing.

I ask, is there any way to live without becoming a subject of electronic writing? Without the presumption of an always missing embodiment? Without capitulating to nothing more than the gesture, the invocation of an elsewhere? Without becoming a subject of, without being subjected to, as the subject is inscribed?

The network is always on. All my dreams are of the net. Awake, I am online in a real virtuality. Online, I am awake in a virtual reality. In front of the computer, the screen blank. How far does the net extend? To ask this is to rhetorically extend and serialize the net's hidden connectivity. The blank display is graphic, graphematic. The display is on. I am on display. Is this not electronic literature? Is this scenic reflexivity and mirroring not the point of our engagement? We give these practices the name writing because they are already written. The question is idiotic and impossible, as are all questions of literature.

I am thinking here of the poetics of electronic literature, of what it invents. Poetics means thresholds and boundaries. Thresholds mean the failure of poetics. Or thresholds of poetics are failure. Think of Jean-Paul Sartre's description of "the poetic world" as "love of the impossible."[5] Think of catastrophe theory, a la Rene Thom: the point is changes of state, intensifications of distended surface. All that is digital, all that is on the net, is this change of state. A catastrophe of writing, writing as the only fact, the fact of inert characters, materialities, transmission, and display.

There is massive production of writing on the net. Call this the literal net, the net of letters. How does the literal net become literary? It forms aggregates, blockages, temporary assemblies, and we recognize these as works with authors, and we discern in the work a certain interest and style. Electronic literature requires aggregates called style, plus systems of knowledge, and ways of handling or comporting the self. Style in Maurice Merleau-Ponty's sense of the chiasm of subjectivity and the world, as the discerning track of the artist, as the expressive actualization of the work, as recognition of the subject.[6]

Style is easily codified. It can be technique. Is your style Java? Flash? Processing? You know how it is. The newest technique. Gotta have it. *That new work was cool, I've got to learn that technique.* Of course, there's genealogical style. Are you langpo? Are you concrete? Are you postmodern? Then there's what holds the style together: signatures, the writer's name. Signatures in archives, such as the Electronic Literature Collection, signatures that register in history texts such as Chris Funkhouser's, signatures in the communicative positions of a theory of authorship such as Philippe Bootz's. Signatures become readable in the work because of these archives. Style is never a trace but a formed work.

What these archival aggregates of style achieve is deepened reflection. Electronic literature is reflexive, aware of influence, aware of citations, aware of techniques. They are a triumph of reflection and a triumph of technique. We post-philosophers of digital text are shrewd and hard-boiled. We believe in a culture built on embedded, imbricated inscription. The belief is so basic that we play out the relation even when it is erased, unreadable. Consider the indifferent blank of code as one possible unreadability, as an epiphany of cultural knowledge. In this literality, we find a deep philosophy of the text as exclusion of the world. Readability is voiding. We can talk of the semantic field of electronic literature, but ultimately our interest is technical. Electronic literature presents the illusion of being untouched by wordliness and thus nothing but wordiness.

Google has style, too. It is ready-made literature. Electronic literature artists make work indistinguishable from Google apps or searches. Maybe you use Google directly in your electronic literature work, or your work coheres as a work much the same way the data midden of Google coheres. Believe me: we all aspire to be Google but are not, or at least hope we are not. Our style gathers as an assertion of being, an aggregation of data. It is the same data as Google, of course. We are happy robots. All the greater need for the signature: to differentiate the work, to assert the occasion and perpetuation of our existence in the work. Is this not the entire point? Is this not the entire enterprise of electronic literature, despite claims that the work is interesting or beautiful? These are surface distractions, and the goal is survival of the writer through the signature, a name declaring that this work is not anonymous, not Google or an app, nor it is a name that owns and archives production. The work survives in relation to the crowd. The writer lives on through electronic literature.

Remember Kenji Siratori? *Blood Electric*, 2002; *(debug.)*, 2004; *Human_Worms*, 2004; *Smart-d*, 2004; *Gimmick*, 2005; *Gene Dub*, 2006; *Non-Existence*, 2006; *Mind Virus*, 2008; *Necrology*, 2010; *Peripheral Psychosis*, 2011. Along with these are various collaborations. Along with these are many pieces of music. Chattering, endless, infected, replicant, overloaded, auto-consumptive, cadaver mechanism, rave texts as "body emulators." The mysterious Siratori: nothing is known of the writer beyond the style and signature, nothing more than texts, nothing beyond, a silence of what he terms HUMANEXIT or "abolition of the world." Nothing but an inscription that fades, downloaded too many times, wiped clean. A virtual idol, like Kyoko Date, or like cutie Hatsune Miku who currently works hard promoting Toyota. Or like Sondheim's Nikuko and Julu. Pathological enunciatory protocols. William Gibson describes the virtual idol as a "desiring machine" where the "only reality is the realm of ongoing serial creation," a "modular array" which "would ideally constitute an architecture of articulated longing."[7] We are all like this: Kenji, but also you and me, are plug and play dub tools for text production, carefully managed presences, arrays of desire, shells of the self (thicker and hardened more than a bubble), that absorb and ward off suffering. Think of the writer at conferences such as the Electronic Literature Organization annual conference or the E-Poetry Festival: there is a description, a narrative, a biography of this subject in terms of works and techniques involved in the works. The subject is clean in these descriptions, scrubbed of fleshiness and mortality, scrubbed of involvement. Electronic literature subjects survive in this way, in an enclave holding in and carrying the writer's cleaned body. My thoughts are impossible and idiotic, as are all questions of poetics.

To exist only if inscribed, archived, and circulated. To exist only if there is a positivity, a remainder online. To exist only if inscribed for others online, and for the sysadmins and roots and NSA who register our being as net traffic. What remainder? Thematics, narrations, fantasies of thresholds, of leaps, allegories of existing online. Delirium of the threshold, delirium as a passion of representation. Sartre wrote of the "mad enterprise of writing in order to be forgiven for my existence."[8] We tell each other, we tell other artists: your work is interesting, it is good, it is beautiful, it is moving. We pass the word to students: check out this work! As if something is there, something beyond the discrete characters and the community around it. Think of ASCII code: nothing but mappings of f(a) to f(b), nothing but sets of functions. The problem of the literal and the literary remains unexplored, or rather unquestioned, since it is all we have: electronic literature as allegory of the literal becoming literary. No doubt there is literature, too much literature on the net. But how? Through what poetics?

We like to discover a voice in code. This seems to prove the well-formedness of the author's survival. We like to talk about embodiment in terms of representation. Posthuman theory deals with embodiment as an instantiatied, historical, and contextual experience that is necessarily performative, and leaves its traces in texts. It is framed for us by discourses of commerce, of power, etc. Only by writing a performative text that speaks as a subject can we exist. Only this writing is available to us, nothing more. Nothing beyond the assertion that this set of codes is mine, I authored it, not you. Nothing but the hoarding of information to form a signature, and to shut out the crowd, the anonymous, and the group. Isn't this what we face? Is not Google a crowd, a clamor? Is there any other existence? Any other way to be online?

Listen close: electronic literature is not an arms race of ever cooler and more refined technique—this is literature as a tool of rhetoric—nor is it pure invention from the symbolic scansion of empty spaces—this is literature as a philosophy of the performative—but it is literature as *elsewhere*. The book beyond the book, the net beyond the net. Nothing to say about electronic literature, and nothing should be said, it says nothing. To describe, to reflect on the system, is to lose the world. Literature means bearing and enduring that ground. The second to last sentence of *Aesthetic Theory* tells of the suffering that is art's "expression and which forms its substance. This suffering is the humane content that unfreedom counterfeits as positivity."[9]

You might reply, look Sandy, you're taking the net itself as proto-inscription, and in such inscription there's no reflection, no intersubjectivity, there's nothing to be read, nothing of interest, nothing at work, and no recognition of the work. As Jacques Lacan put it, "If I press an electric button and a light goes on, there is a response only to my desire."[10] So, you might say, Sandy, you're just intensifying, lighting up our desiring production in and of the net. You might add, for the becoming-literary of the literal, for the peeling away of the character into delirious allegories that are the phantasmatic ground of electronic literature, *for all this*, there must be otherness at work, an other to whom I write, who reads, who lurks on the net.

I reply, it is not otherness but intensity, suffering, and the ecstatic surface of the world. My body is marked. It is not coded, in the sense of a repeatable code approved by sysadmins. It is not a performative act to be recognized by the crowd, the community. It is a display, an ostension. Think of ostension as showing, as making manifest, as a strategy of mimetic pointing. With this, a logic of ostentation: showing off, showing too much, a display that mimes and spreads. Ostentation is splendor and also suffering. We bear it. And finally it is a portent, a pointing toward the future. Display as ground of electronic literature is the temporal future anterior online.

Murmuring of the world as noise and visibility, as the coherence and condition of the net. The net, not as a protective shell but a consumption and inhabitation. It feeds me, keeps me alive. The network is on, an extended membrane, a fragile skin, a marked body. Afghanistan, Somalia, Tibet, Rwanda, Crimea, Detroit, West Virginia coalfields, West Side Buffalo.

The crowd of electronic writers

Where is the literary community?[11] Everywhere, as are the works of electronic literature, everywhere, in every medium, on every surface. Media's indifference to what it signifies, indifference to what it performs, and indifference to materiality are all enabling conditions for literary community.

There is literary community and there is the crowd of electronic writers. The crowd is mobile, with skilled access to capital and to practical techniques. Within this crowd, we have nothing in common. I share nothing with you, and yet here we are. Such is the condition of our relation and enrollment in the organization: there is no community here. No, in the crowd there is multiplicity.

Crowds are constituted through and operate on surfaces. Electronic writers are a crowd operating on the surface of electronic media. What constitutes and operates the crowd?

The crowd of electronic writers can form around use of technologies. Look at the crowd of writers using Flash or the crowd using iPads. They crowd around technologies, they operate through technologies, and they recognize themselves as a crowd through this operation. Crowds can also form around other crowds. Look at a crowd of electronic writers amidst the crowd of new media artists or amidst digital humanists or amidst PERL programmers. They are in a crowd of other inventors and creators. They crowd on the surface of media. They recognize themselves through their use of language and through their identity as writers. Crowds can also form through codified communications. Look at crowds emerging from listservs or blogs. They crowd around the site of communication, and they recognize themselves through their repeated inscription in the site. Crowds can also form with pressure from above through availability of funding and prizes, university positions and residencies, and the like. Look at the crowds following electronic literature research projects such as ELMCIP[12] or following conferences such as the Electronic Literature Organization conference. The ELO is nothing if not a crowd. They, or you, or us, crowd around conferences, crowd around readings, and recognize themselves (yourselves, ourselves) by being at these sites and by participating in the themes and debates held there (or here).

In short, there is a crowd of electronic writers. Is there a community of electronic writers?

Crowds differ from communities. Communities of friends are present through hands for shaking, voices for mingling. My child cries and I commune with her, sharing and soothing. Communities are organized on the skin and within the body. Amish or Mennonite communities dress simply, cut their hair, and hold their bodies in specific ways and according to set protocols. The Ndebele of Zimbabwe wear neck rings, force their necks to stretch and lengthen and accommodate more rings, all signs of belonging to their people, to their community. My daughter pierced her ear and then pierced it again and then pierced it again, all in the same area, all the same ear, and this scarring and opening on the surface of her body marked her as a member of her community. The Ndebele's neck or my daughter's ear or the clothed Amish feels the community on the skin and within the body. Communities are organized by marking bodies, by marks readable only by those in the community.

Communities of lovers touch bodies, share warmth, and offer each other stroking fingers, smeared fluids, and soft and hard membranes. There is no address or reference in communities of lovers. Their discourse is impossible, meant only for themselves. It is a discourse of "extreme solitude," as Roland Barthes put it, and yet also of affirmation.[13] How do you read the words of lovers? I do not mean how do you read the words of your lover, the one whose words are addressed to you, but how do you read from outside the words of a community of lovers? How do you observe lovers and read off their community? Such observations and such readings recognize the impossible secret that is community, the love and hate that it involves. It recognizes this secret through the emptying and distancing of discourse, in short, through impossibility. It is impossible to read the words of a community of lovers and instead we read only the mystery and privacy of their intimacy. No writing is sufficient for the love of lovers, which is why there is no end of writing on love. In short, nothing marks the lovers; they are marked for each other. Lovers know each other, read each other, but there are no writings that show their community. They are a community.

Crowds are not communities. The crowd of electronic writers is well articulated, well formed, but is not a community. The crowd is constituted and operates first and last through inscription on surfaces of media. The bodies of electronic writers share nothing. It is not that we are not embodied; it is not that we do not bring our bodies; it is not that our bodies are not here; all of these are the case but none of these are what constitutes us as a crowd. It is our inscription on the surface of media that constitutes us as a crowd, not the intimate sharing of community. I share nothing with you, we have nothing in common. We are no community, we are no we. There is not a community of electronic writers. Yet we all may be lovers of literature, we all may be members of the literary community.

The electronic writing crowd does not start from the mingling and presentations of organs and body parts. Not at all: it starts from the non-presentation, from the negation and gap across which no bodies are presented. Non-presentation, negation, gap: the crowd starts from attachment to a medium, to a tool, to an operating system, without reference to bodies or intimacies. Put aside your body and your self, in order to be constituted and operate as one of the crowd of electronic writers. The crowd is a commitment, a contract, and a membership. You join the crowd just as you join the Electronic Literature Organization.

The crowd may include those already in community with each other. Lovers and friends in the crowd share with each other. I think of Christopher Funkhouser, a scholar and artist working with electronic writing, I think of him as one of my oldest friends, someone I met over twenty years ago. The day we met was before either of us was part of this crowd. Neither of us was involved with electronic literature. I remember Chris telling me that no poet would ever write on a computer. The two of us mingled, followed paths, collaborated and communed, and here we are. I feel Chris and I share and are in community, but that is fundamentally different from the electronic writing crowd we enter into. I write a blurb for his book, he writes a letter for my tenure: all this is part of our work in the crowd. Even if there is community in the crowd of electronic writers, it does not make electronic writers a community. You may be my lover, we may share ecstasy and intimacy, but that is never the commitment and condition of joining the crowd of electronic writers.

You see what this means, do you not? I create or write about works of electronic literature built on indifferent drives that float and skew across surfaces of media. The writing is on the surface for the crowd to read and write. It is intractable and real, yet distant, an other work that remains intimate and secret. In the ecstasy of media, I dream this other work. I am the subject of electronic literature. I am where the work is realized.

You might respond and tell me you share all the time. You share stories on Facebook, you share music through file torrenting, and you share data through USB or memory stick or Dropboxes. I might reply that there is sharing and there is sharing. In these examples, it is not the sharing of bodies but of codes. There is intensity in code. I take pleasure in codes that display and assert you, I take pleasure in your Facebook posts and the files you give me, I take pleasure in your images, in your texts. I feel pleasure in relation to the code. And not just pleasure: I also suffer in those codes that display and assert those that suffer or those that threaten me. A webpage that displays pain or atrocity or injustice causes me to suffer.

Unlike the immediacy and contact of community, the intensity of code is a delirium of the other's imagined body. I decode you through shared files. Such is the mingling of the crowd: our files share and we read each other in and through the files. Such sharing of code is not about bodies, not about intimacy but about imagination, which is its power. The code is on the surface of media. It is the negation of bodies. Your Facebook page and your driver's license are abstractions in relation to your body and your concrete existence. They are files that circulate and allow you to share in the delirium of the surface of media. Your intimacy and privacy is a setting, just as Facebook constantly reminds you to check and set your privacy. These sharings must not be understood in terms of the intimacy of bodies in community but in an encompassing capital investment in floating images through axiomatics of media, around which form varieties of crowds. The crowd's delirium is the absence of mingled bodies. The crowd's mobility and speed is the lack of touch, is the hollowing out of organs, speeding across the gaps between skin and skin.

The term "community" is used in many ways, including to capture the fantasy of sharing achieved by the crowd. I am attempting clarity in using the term "crowd" to recognize that the crowd—not the community—of electronic writers is constituted on the surface of media. What difference does it make? Multiple surfaces are available for the crowd of electronic writers. In fact, unlike the community, with its depth and commitment to bodies, the crowd is mobile, growing, and intensive.

The necks of the Ndebele or my daughter's earlobes are singular and attached to bodies. They are part objects in close proximity and never straying from interiority. The crowd is always straying, always detaching, and always mobile. We are at the periphery, riding the edge of the crowd. The crowd of electronic writers can mark iPhones, Twitter feeds, Arduino controllers, an electronic billboard in Sao Paulo, a paper that prints out computer code in Cambridge, a screenshot on my computer at home, a webpage, and so on; the list is infinite. Any surface will do, any sign.

Call this the *designative* function of the surface of media. We will attach to any flow at all. Our crowd is indifferent to the marks and means of signification. We are not writers in the sense that we do not work directly with words. We are writers in another sense: through the contractual logic of the crowd, anyone writing on any surface of media can run with the crowd. The electronic writing crowd can operate in ways community cannot.

Debts and obligations

Preta in Sanskrit or "hungry ghosts" of Chinese Buddhism are unsatisfied and desiring souls, partly alive and nourished through scraps. I am hungry. I am frantic to maintain accounts. I keep up my home page, archive documents, remind others—you—and myself of my existence. I post to Facebook and check email. I fulfill the debt and obligation to be myself online.

The crowd of electronic writers is no different from any other crowd constituted on the surface of media, in the space of flows, and in relation to capital. This crowd is no different from the crowd of net art or interactive fiction, but also no different from the crowd of trout fishers and basketball players, no different from crowds of plumbers and nuclear physicists, no different from any and all other crowds. The dynamics of crowds operate regardless of content and concerns. So much of what is done in the name of electronic writing is explication and application of this dynamic. So much of what is written and researched about electronic writing is insistence that the crowd of electronic writers behaves like all other crowds. Perhaps we do behave in this way, but perhaps we do not.

Rather than marks or signs, I tell you that writing is permission and obligation. The trajectory of the writer is one of commitments and tributes. What obligations? To history for a start: the history of writing and the history of literature. We all feel the obligation, we all tally the account. Electronic writing owes a debt to the avant-garde. Electronic writing owes a debt to concrete poetry. Electronic writing owes a debt to poststructuralist theory. To OuLiPo. To Language poetry. To postmodern fiction. To Alan Turing. To Ada Lovelace. The list continues. We pay homage: to talk of electronic literature is to give examples, to reference other writers, to insert myself into networks of circular debt. In addition: electronic literature owes debts to the future, to emerging writers, to the need to keep up to date. Like all electronic writers, I feel obliged to know the latest work; I am in debt to the emerging future of electronic literature.

Debts are tallied and the accounts are balanced through scholarly and critical practices around electronic writing. Tagging, indexing, and cataloging: these ring up what is owed. ELMCIP project or the Electronic Literature Organization or any other example you choose are about the constitution of electronic literary works through obligations and affiliations. They are about an investment in maintaining the author's name and the work's name and the debts surrounding these names. Such projects claim we need to form communities, we need to identify works, and we need to enable artists. My point is not whether this is true or false—it is good enough, it helps us out—but rather to point to the debt assumed, to point to the enacting of obligations, and—with this—to point to the way this claim to community wants to please, wants the pleasure of being stroked by the other.

Most of all, there is a debt to the machine, a debt to software and devices, debt to the mega-apparatus that is the network. I work on a computer that cost over a thousand dollars, and I use devices costing millions of dollars in research and development, and I communicate using telecommunications systems costing and generating billions of dollars worldwide. Writing rides and expresses capital's flexibility and extent. It is not possible to be a writer without assuming and expressing this debt. It is not possible—for anyone, in any context, at any time—to discuss electronic literature without re-announcing this debt. Debt is the work. It is the condition of writing and the discourse it produces. The Electronic Literature Organization's definition of electronic literature as "works with important literary aspects that take advantage of the capabilities and contexts provided by the stand-alone or networked computer" is a way of saying that such works are obligated and in debt.[14] The work "takes advantage" of what is provided, and only becomes the work, only becomes literature, by this taking and by this provision, by the part that is given, and by the advantage that is taken. Writers continually accumulate and pay off the debt through the creation of work. Is there an exit strategy? Can we refuse our obligations?

You might ask, why debt? Why not generosity and infinite gift giving, why not writing as a poetics that always invents anew? To which I reply, look around you. We occupy cultures of debt, of borrowing, of mortgaging off our houses and jobs. How odd it is to think of such conditions separate from and not applying to our writing. Electronic writing is possible only if it starts from the condition of debt. The crowd is a complex sociality—it is not the immediacy and intimacy of bodies—and its density is floating commitments and protocols. Of course, writing is both contracting into and breaking up contracts. Debt may be paid or not. To write is to parasite institution, to occupy its space, to set up on the medium and generate the work.

Literary community is not fulfilling obligation and debt, but acceleration and release: acceleration of myself, release toward the work. Literature breaks all debts and obligations. To write a work of literature is to write such a break.

I may never break through, I may never write literature. I dream of breaking through, I dream of writing literature.

Axiomatics

All these are axiomatics, self-evident operations of media on generalized flows in differential relations—think of Castell's "space of flows"—without specifying specific media forms or media specificity or encodings. Debt and obligation are a generalized way of understanding the axiomatic relation to media, assuming no prior condition: no format, no materiality, also no memory, no reference to truth or falsity, only combinations and productions. The point is not codes or significations—rather, codes and significations become the message or content of the media—the point is rather to institute a "substance" that can be worked over and made into works.

The intensity begins from the writer's name. The name can be any inscription at all. Intensify any inscription to make it a name. All the questions of reference, of truth, of rhetoric are used up and no longer relevant. The surface of medium is not a sign but a splitting and an elsewhere that organizes signs around it. So too the electronic writers' name (and the works that are given the writers' name) are references to the surface of media and not to a putative biographical writer. This is the case even when the biographical writer is present—as many such writers often are at conferences or readings. In such cases, the writers still defer to the medium, they are there to show the work, and they are there to describe its construction. Their presence, their body, or their interiority is secondary and incorporated into the relation to the surface of the medium.

What does electronic writing mean? Can we even ask this question? Refuse it. Instead, start from points of expression, start from the extended surface of the medium as cybernetic zero point. The electronic writer dreams of giving a name to the zero.

Zero byte files are computer files with no data. Such files are nevertheless still image or text files, and are designated by a.jpg or.doc or similar naming convention. They are files produced or read by an application. Yet they contain no data. Zero byte files are named and therefore stored on the computer. They are nothing but the name.

Crowds constitute and program themselves through the commitments and contracts, and not through the communion of bodies and parts. Crowd operates through mnemotechnic delirium rather than bodily intensity. Because crowds come with nothing, because the axiomatics of media negate all memory and traces, the crowd must create memory for itself. The crowd's memory is always artificial. It is any memory at all. The point is not a particular memory but mnemotechnic production from the cybernetic zero point of inscription. The crowd operates as if memory were attachment, as if it were the navel binding all to the medium. It is the admirable work of organizations such as the ELO and projects such as ELMCIP to construct and narrate this memory.

Heinz von Foerster wrote of "zero order cybernetics," the autopoietic threshold "when activity becomes structured; when behavior emerges, but one doesn't reflect on the 'why' and the 'how' of this behavior. One just acts." He concluded, "This is when cybernetics is implicit."[15] The implicit zero point is what we remember after the fact. It is the navel, where we constitute ourselves in relation to the infinitely stretched thin medium that is everywhere.

The dream grows to sharing the name and what it includes. The fascination with text generation, for example—such as Nick Montfort's *Taroko Gorge*—cannot be separated from the act of naming the flow of generated text, naming it as mine. The point is not to approach the generated text as meaningful or not; instead the text becomes interesting and significant because of the author's name. The name illuminates the text. The nonsense of the illuminated text indexes the naming of the zero point. We discover structure, movement, and meaning in the generated text precisely because of the function of the name.

Electronic writers enter the crowd by naming their work electronic literature. The name is the latency of the crowd of electronic writers in any and all medial flows. It is as I said already. Crowds form with pressure from above through availability of funding and prizes and university positions and residencies and the like. These names and institutions situate the crowd, hold bodies in place, and provide a name for all to share.[16]

Or, crowds form through codified communications, emerging from listservs or blogs, crowding around the site of communication, and recognize themselves through their repeated inscription in this site. The code of communication is the memory they share. They name themselves in this way.

Or, crowds form around other crowds. Think of the crowd of E-Poets, identified in relation to avant-garde poetry, or the crowd of experimental pomo fiction writers now become electronic literature writers.

Or, perhaps easiest and most common, crowds form around use of technologies. I am a Flash poet, I can say, or I am a code worker. I can present my output, I can name my work.

Think of Gregory Bateson's "plateaus," later adapted by Gilles Deleuze and Félix Guattari, defined as "some sort of continuing plateau of intensity [...] substituted for climax."[17] Bateson draws comparisons to Balinese culture, to trance states, to the infant held in suspension by the mother's erotic stroking, to legal quarrels, to music, and to warfare. There is refusal of termination, turning away from fixity, and withdrawal from conclusion. There is no end, no goal, only continuing intensity. The emergence from cybernetic zero point to the mnemotechnics of the name establishes a plateau for the crowd, a common intensity felt within our bodies but not shared communally. Each electronic writer occupies the plateau and is able to announce: I am a writer, I belong with and follow the crowd.

Plateaus of intensity both negate and substitute for the absent body. Writing is the plateau. It is insufficient to understand writing as manipulation of material traces, even in the modified form of matter organized through digital codes. Such a view is an atavistic hangover of the content and substance of earlier situations. We are not writers in this way. The axiomatics of media are immaterial and virtual: any and all markings, whether digital or not, whether on the net or not, can be taken in to the field of electronic writing.

The literary community

It is obvious—though no one seems to be willing to write this down, to admit to it—that no theory exists to analyze literary texts and signs on the computer and the network. There is almost no consideration of electronic writing as literature. We still do not know how to think about the literariness of media. What is literature and how is it possible? Surely this question—maintaining it, following it, keeping the question alive—is what fascinates us and why we continue to discover literature as this very question.

Do not get me wrong: there are admirable descriptive formalisms and historical genealogies of electronic literature. All these function as criticism should, but they do not explain how such inscriptions become or exist as literary. To describe genealogies—I think of Christopher Funkhouser's admirable work or the community mapping of the CELL project—is not to understand conditions of existence. I wish I could say that we tried and failed to produce such a theory, but I cannot. We have not even tried. In fact, there is no attempt to determine the necessity and existence of electronic literature. Our dealings with electronic literature so far—again, not critical descriptions and genealogies but fundamental philosophies—are meager, amounting to little more than repetitions of early literary criticism.

At stake is whether we can delimit a domain of electronic literature separate from the totality of literature (literally, all written texts). The problem is that existing criticism begins from the presumption that "there is" electronic literature and proceeds to describe the various works in existence. In *Electronic Literature: New Horizons for the Literary*, still the only book explicitly on the subject of electronic literature—that is, on literature in the large sense of the excessive mass of produced texts—N. Katherine Hayles explicitly refuses to theorize the subject of her book. It is not the place to do so, she declares. No doubt the results are productive for maintaining distributions of texts and readings in a field of literary and non-literary inscriptions. No doubt, as well, the results are useless for a theory of electronic literature.[18]

So far I used the term "literature" sparingly and only to refer to the works that fall within the circuits of debt and obligation. I refused to refer to electronic literature in any general sense because

I was describing the crowd of electronic writers. There is, on the other hand, the literary community that exists not because of the dynamics of crowds that characterize capital and inscription in relation to the surface of media but in and through the shared intimacy that is literature.

To read and write literature is to enter a secret community of lovers, lovers of literature who are able to share in writing and reading but unable to say the secret that brings them together.

The poet Robert Duncan wrote of the "drama of our time" as the coming of us all into one fate, "the dream of everyone, everywhere."[19] Dream here does not mean something insubstantial and passing in the night, or something standing in for a personal psychodrama. It means precisely a delirium of sensation because of a relation to the surface of the medium. The delirium is global, geopolitical, racial, and gendered. It is delirium deliriously quoting cultural material. The delirium of the medium lets me dream of the other's body—dream of you—in emails or txts, in photos, in Skype calls. In all these, I recognize you. The "no body" that situates my body elsewhere leads me to imagine bodies in every flow of text, leads me to imagine displays of your body in every image.

No flesh is bared. No organ is presented in electronic writing. No flow of text is bodily flows of blood or mucus or saliva. My body is bound and fixed before the computer. Yours is the same. No trace carries over, no body part, no fluid, no piece of bone. The fact is axiomatic, material, physical, and philosophical. The computer's mouse may measure the movements of my hand but it is first and foremost an object of breaking and sending. In the absence of the body, in the negation of the real, sending is infinite and breaking is intense. Stopped and coupled, I display my dorsal plane, engage in ape-like flashing of my posterior while I flow and seep across the machine's surface, while I lock and load onto its display.

The dorsal-ventral axis or DV connects the tips of my body. Top of the head to feet, buttocks to belly, with semi-symmetrical halves on the left and right of this seam. The rostral from Latin *rostrum* or cranial from the Greek for skull or cephalic from the Greek for head refers to the forward angled head of those beings with distinct heads on their anterior; that is, us. Our bodies are cephalic and ventral. Ventral from the Latin *venter* for belly. Our bodies are angled forward and upwards. Our ventral surface, defined by the seam of the DV, opens us to the world, or closes as we close our eyes and fold our arms and shoulders inward on the axis. The ventral is our facing and direction. The coronal plane divides my body into ventral and dorsal, slicing front and back, transecting my standing body in half. We are crossed by lines and dotted by points. Upright and on display, we project a transverse plane parallel to the ground and forward from the ventral axis toward the world.

The ventral and transverse project from my body. The ventral and traverse are stopped, put into relation and transformed, the axes segmented and the plane broken. What describes these changes? Stoppages: the eye's gaze stops on the thickness of things, the skin on the surface of my finger is stopped, and the flow of saliva or mucus or semen or blood is stopped. Stoppages by things encountered, stimulations and intensifications at edges and openings. Stoppages for upright beings, cranial and cephalic organisms who face objects above the transversal. Stoppages or scintillations: the flashing display of our surface, continuous posturing and display of our own bodies, offered to the world with no hope of return. A computer screen is for bodies organized like me and for my kind of surface. The screen stops me. The screen couples with the ventral and transversal.

The intensification and sending of the stoppage means forget your name, forget your files, forget your works. Literature blurs and vanishes all of it. Literature discovers organs sprouting from the medium like flowers. They are the sounds and the visions that let us speak and let us see and hear: images on screen, sounds, and words. There are organs everywhere on media—wagging, jerking, and splayed. The place of organs and body parts and objects and detritus is released into the abstraction of the medium. The flows were mine, were parts of my body's becoming liquid. The flows on the surface of media are abstract, are nothing but flow. This de-subjectification and un-signing of flows means that all the bare flesh on the Internet—from Facebook faces to endless porn—is not bared and reveals nothing. They are cool or lukewarm image-codes that we cannot evaluate. How is it that I come to desire the array of pixels in a porn image or identify a Facebook face as that of my child?

VI

I read my spam

*every living things does everything it
can, not to preserve itself but to become* more.

NIETZSCHE[1]

Spam is a proto-textual and pre-broken domain, wounded by the body, broken by voice. I read my spam and want to respond, to correspond with my spammers. Some recent spam in my inbox: "one must speak a little, you know. it would look odd to be entirely silent for half bennet, i am inclined to think that her own disposition must be naturally bad, or she could not be bingley expressed great pleasure in the certainty of seeing elizabeth again." You recognize this right away as fragmented and re-processed Jane Austen, hidden within advertising for a stock trading website. My email client filters out html and images, leaving only these cryptic writings.

Spam texts are encoded but no decryption is possible. There is no plaintext message. I find them wonderful, and read them as poetics, as odd fragments generative of narratives and scenography. I find the process of their production wonderful as well. The texts are written to elude community standards and means of censorship, and at the same time to enter and impose themselves into the standards and means for the community to read itself. Spam text attempts to make itself equal to a constant murmur and flow of communication that we all receive, as an imaginary and unwritten *writing emission*. Transmitted only because it breaks its own transmission, produced only because it collapses its own production, spam purposely enters into a pre-broken domain of *smeared text*.

Messages like this come to each of us. We are all recipients. There is a hystericization of spam, leading to immediate armoring and closing down of ports: we clamor for protection from this threat, which some claim leads to more than $20 billion in lost productivity through time wasted and servers clogged. Our individual irritation with spam is the correlate of this economic logic. There are exceptions, aberrant opinions, such as Willard McCarty's speculation as to whether spam could "yield breakthroughs in text-analysis?"[2] If it is typically approached as a matter of security, of hiding and concealment, it is because spam is inseparable from the literary problematic of the web, a problematic of text that neither reveals nor conceals but gives a sign, as Heraclitus said of the oracle of Delphi. Spam is tied to the psychoanalytics of display and concealment, and to the absent body.

The spam messages I quoted are specifically composed to elude the Markov or Bayesian filtering techniques common to many browsers and email systems. Spammers copy the passages from online texts to increase the overall improbability of the writing in the email. The added passages shift the probable occurrence of the entire text away from the probability that it is spam. In comparison, strict keyword filters flag email because it contains certain words. A filter may tag the word "penis," for example, but I may want to send email with that word or with the phrase "refinance your mortgage." Moreover, simple substitutions (such as the leet of "p3n1s") elude keyword filters. Bayesian filters, a statistical method dating back to the eighteenth century, measure the conditional probability that an email text is spam. They evaluate email text in relation to the probability of the text occurring in any email, and against the probability that any email is spam. Is a word or phrase probably spam in general? Secondly, is it probably spam within the semantic domain under consideration? Bayesian filters must be "trained" in the probabilities of a given system, a user noting which writing is spam and which is not, the filter gradually "learning" whether I frequently use the word "viagra" in my email, and so on. Over time, the spam probability of a given text is computed in relation to all text. Writing that exceeds the spam probability threshold is filtered out.

As a result, all text is probable spam as a condition of its communicability. All text is an amalgam of spam and communication. The intention of a text, one directed to me and that I desire to read, is always partially spam. The political economy underlying spam renders text as product, and renders my reading of the spam as a medium or interface to consumption. In this sense, spam is a perfect model of the communication circuit, of the message received. The discursive field around spam is larger than individual messages. The goal of spam may not be reading and clicking at all, but seeding phrases and URLs into the dynamic total sememe of the web. Perhaps spam is not about getting people to look at ads but to push certain texts higher in search engines such as Google, where fields of interconnected terms and links determine page ranking. Search engine optimization is the "white hat" version of spamdexing. In turn, the recent counter-spammer technique of chongq-ing relies not on filtering or denial of service attacks on spam sites but on creating links from spammed terms to non-spam sites to fight the spammers at the level of search engine ranking. The result is a struggle over the link-based "value" of phrases in the textual sememe, a struggle in and through writing, a struggle amplified by the distributed dynamics of the so-called Web 2.0 of wikis and social networking software.

Spam theory collapses if it remains within the capitalization of exchange and circulation. The spammers who write me insert passages to displace the overall probability of the message as message. They create noise in their own text that disturbs its identification as spam, whether in reading or in spamdexing. Oddly but wonderfully, spam texts become a writing that tends toward the conditional, the heteroglossic,[3] and toward the unfinished. The spam passages are often from literary texts, often mined from Project Gutenberg and other free sources.

More than this: the spam does not simply quote literature but tends toward the literary. To bypass the trained Bayesian filter, spam must singularize itself, must render itself improbable in any written environment. Spamming is a poetics of transitive writing, a writing toward its own erasure and silencing as communication, silenced toward its own erasure as spam. To escape the filters, spam is written toward the singular and imagined other. I read my spam and want to respond because of the otherness of a communication that is written to me, singularly to me beyond every possible statistical measure. I receive the text—this from a spam for some sort of energy stocks—and I open it, and it is meant for me: "A stovepipe defined by a sandwich takes a peek at a South American ski lodge.

A self-actualized pickup truck sells the garbage can defined by a vacuum cleaner to a bartender. Some precise food stamp conquers the diskette. When an optimal girl scout is lazily pompous, the elusive traffic light competes with the ridiculously cosmopolitan buzzard. A satellite eagerly eats a cyprus mulch."

PLEASE REPLY MY BELOVED

In all cases the physical occurrence of an index word is bodily annexed to what the word indicates. Hence "you" is not a queer name that I and others sometimes give you; it is an index word, which, in its particular conversational setting, indicates to you just who it is to whom I am addressing my remarks. "I" is not an extra name for an extra being; it indicates, when I say or write it, the same individual who can also be addressed by the proper name...

Gilbert Ryle[4]

"My name is Mrs. Mellisa Lewis," reads the email. "I am 59 years old and I was diagnosed for cancer for about 2 years ago. I will be going in for an operation later today." Bad for her, but good for me because she is going to will me "Fourteen Millions Two Hundred Fifty Eight Thousand United States Dollars" for "the good work of the lord." She provides instructions on how to claim the money. She ends the email asking me to pray for her recovery and signs it with regards.

Mrs Kim Paul emails me later on the same day. The subject of her letter is "PLEASE REPLY MY BELOVED." She too is willing me her estate. Luda Johnson emails the next day. Puzzling to me, she came across my email address through "an email surfing Affiliated with the US chamber of Commerce." What does that mean? She offers a female puppy for adoption. Will I care for and show an interest in the puppy?

Someone calling himself Patrick Dooley emails: "Hi, I hate to be the one to mention this, but people continue to talk about your weight issue and it just disgusts me." This is also puzzling, since I did not realize I had a weight issue. He continues, "Whether you know it by now, people are always chattering about each other at work but you come up more than enough." The news is a complete surprise, but it turns out he has some "stuff" to help me, available on his "anonymous email website." He instructs me that "When it helps/works just send a memo out with the name 'Angel' in it. Then you can take me out to lunch to thank you." Amazing! But why a memo with the name Angel? Later, Vivian Dabah, a 23-year-old girl from the Sudan, emails. She wants to get to know me. She wants my understanding and interest. "Take care as I wait for your response through my private email address above," and she signs it "Yours, Vivian."

It is easy to laugh at the simplicity of these emails. They are ploys. Such laughter is about distance and control. I am not touched by the email's call. It addresses me but I know what is at work there. I know the type: the spammer, the con artist, and the criminal. I possess and employ knowledge of the otherness at work within and behind the address of these emails. The lure or call of spam is purely a "trick," as social media theorist Clay Shirky explains.[5] He intends to be reassuring, to contain the trick within a repertoire of social relations, yet such contextualizing highlights and doesn't explain the lure. I am still left with the address itself: how is it that the spam calls to me, in all its inauthenticity?

This question is not dealt with so easily. In a *New York Times* interview, Sherry Turkle identified the fantasy of escaping the burden of email. Declaring "email bankruptcy" is the public declaration of inability to respond to your email.[6] You are addressed and you cannot respond. Your email account is clogged with garbage. You are so far behind you give up on the account and start a new one. The problem may be spam and may also be a surfeit of legitimate messages—too many to answer!—but surely the burden is at least in part that each email addresses us, calls us, beckons us, and the fantasy Turkle diagnoses is of blocking our ears and not hearing the call, not hearing the other through the net.

If you knew how to listen, you would understand that I am talking about a problem of knowledge, of the status and production of knowledge, and the topography that situates and is situated by this production.

It is easy to be cynical. Of course, we hear about people taken in by these letters, newbies and rubes who give their retirement to Mrs Kim Paul or Vivian Dabah. The knowledge I am talking about here is double-sided: on the one hand, the structure of address and its pure convention in email, where PLEASE REPLY MY BELOVED becomes a citation of one possible affectionate formal closing of a letter; on the other hand, the successful performance of this convention that draws in the reader and seduces in its call. The address swarms over me in an instant. It is the highest of high-speed net writing. The confidence trickster plays on my compassion—my desire to help, my desire to respond to MY BELOVED—a trickster playing my greed to get my hands on the "Fourteen Millions Two Hundred Fifty Eight Thousand United States Dollars" that Mrs Mellisa Lewis is willing to me (and which she surely doesn't need, now that she's diagnosed with cancer).

All in a day's spam.

You know the arguments. Spam is offensive. Spam wastes time. Systems administrators call it "unsolicited bulk email." It clogs the inbox. Estimates claim spam comprises 80 to 85 percent of the entire world's email. This is why we need filters, buffers, programs that protect us. The admins approach spam as a problem of resources. If this is true, it is because the system is intended for messages, intended for delivery, intended for addressing. The problem of address is folded into the resources of the system. It is not simply a problem of bulk. It is not simply a matter of spam clogging our inbox or of annoyance and distraction. No, it is tied to who we are on the net as receivers and readers, senders and writers.

Can spam

All net writing is spam of one sort or another. The writing we receive and is meant for us is simply spam that we accept. Who is it that distinguishes spam from "authentic text"? Literary critics, of course. Surely you recognize that this is a distinction based on an aesthetic taste for the authentic text of a subject? Are we not all trained from early on for these distinctions? The sysadmins and programmers and app developers and hackers as well: all care for the problem of authentic text. To unburden and unblock yourself of this vision is to exit into the wild zone of spam theory and electronic literature where every net writing is spam of some sort, and all such writing poses the problem of an emergent murmuring ghostly voice of the poet. I propose: I know there is electronic literature—know as in understand, perceive, and read—because of the problem of literature. This proposal is spammy. This proposal is philosophical. Literature affirms and creates. "There is" literature.

I might invoke the longer history of junk mail. Matthew Sweet's fascinating *Inventing the Victorians* suggests that the Victorians invented spam junk mail.[7] The manipulation of address and the forcing of delivery is not a function of electronic mail but of the postal system as such. I might link to the scam, the grift, the con, the bunko, and the flimflam. Think of the "Spanish Prisoner," that good old confidence trick of luring a mark into giving money to save a prisoner in Spain. It requires developing the mark's trust and sympathy for the prisoner, as well as the hopes of a great reward. Or think back even farther in history to the "Letter from Jerusalem" fraud described by Vidocq in the eighteenth century, where the scammer shows the mark a letter from a rich man in Jerusalem offering to share wealth if the mark sends funds to enable the man's release from imprisonment. The formula is clear.

The primary law covering spam is the CAN-SPAM Act of 2003 (Controlling the Assault of Non-Solicited Pornography and Marketing), signed into law by George Bush. CAN-SPAM was the first law regulating Internet spam. It legalizes most email spam. It focuses on intent. It focuses on spam as email which functions as advertisement or attempts to sell a service. Spam is defined as a "solicitation." If the subject line or the content of the message is informational, the email is considered transactional and not a solicitation. Compare the Spamhaus definition. The Spamhaus project is a major source on the web for spam information. The project's technical definition of spam goes like this:

> An electronic message is "spam" if (A) the recipient's personal identity and context are irrelevant because the message is equally applicable to many other potential recipients; AND (B) the recipient has not verifiably granted deliberate, explicit, and still-revocable permission for it to be sent.

Also:

> Unsolicited means that the Recipient has not granted verifiable permission for the message to be sent. Bulk means that the message is sent as part of a larger collection of messages, all having substantively identical content.

Finally, "Spam is an issue about consent, not content."[8]

In all this, we are dealing with senders and addressees as positions of information circulation in a communicational circuit. It is easy enough to treat email as packaged information, as part of a discursive field where sign production is economically regulated. What of consent (and not content)? Consent is the permission given by bodies entering into such systematic communicational relations. As Winograd and Flores showed, following on Maturana's work on autopoeitic systems, the computer as technical object can only be understood within a consensual domain of human practices and encounters.

I think of such consensus when they email and call me "Dear One" or "Beloved" and beg my reply. Their cancer and agony are displayed in writing. By "display" I mean words that are immediate, that cannot be contained and reduced to the economies of the information circuit. I mean phrases that touch and wound and caress me. Is not the phrase "I was diagnosed with cancer" as piercing as "Beloved" or "dear one?" Is it not as moving as "donate to Haitian relief?" Does it not tug at me the same? How can I address this pull? I insist, this is a problem of address itself, a problem of the structure of being addressed and the convention of address in a letter. No doubt there is the history of this convention. The address is a sign that signifies various formal attitudes. PLEASE REPLY MY BELOVED is a convention between lovers and those in close affection. I know this, which is to say that there is an epistemological attitude toward spam. I know that it invokes these conventions and this knowledge armors me against the spam. I mean armor in a very literal sense: my body is firm, unyielding to this writing. I don Wilhelm Reich's "character armor."[9] I know who I am; I wall myself against the other. Are we not dealing with a political economy of sentiment? So much invested here, so much conserved there. We are armored against the call through the molar arrangement of the subject that we are. Molar subjectivity tied to a sentence: I will not respond, I am not your beloved, you are not an object of my desire, you are not an object of my sympathies. The armor serves an economic function of controlling our investments.

Think of an email address such as vivian1985dh, from a popular email service, or charles.baldwin@mail.wvu.edu. The address is a code that email programs take and use to direct the message across the net. The address is composed of a user name and a domain name. The user name is set by the sysadmins on a particular network, perhaps following a decision by the user, and the domain name is set by the name of the network. The domain name is assigned to particular IP addresses, sets of number corresponding to the Internet Protocol address of a site, assigned by ICANN, the international agency regulating Internet addresses and names. The email address is attached to a message. The address accompanies the message. In what way? Is it part of the message or is it something separate and other from the message?

This poetic problem of the limits of the text, and of ending and beginning, is evident in Crocker's RFC 822 "Standard for the Format of ARPA Internet Text Messages" of 1982, the core Internet Protocol defining email messages.[10] The Internet Engineering Task Force (IETF) Request for Comments (RFC) are basic documents outlining and regulating the function of the net. The important distinction made in this RFC is the separation of the email message from the header fields. While the rigidly formatted and syntactically determined header fields are crucial for transmission, the content is irrelevant and is simply transmitted. Yet the entire message, in its transmittable form, is nothing more than a string of ASCII characters punctuated into fields by special escape characters such as nulls or CRs. There is no material difference between header and content, only a difference designated within the ASCII characters. The message is structured and formatted only at the point of reception, where a reader encounters the problems of the space of the text and the address involved therein.

The end of spam

Giorgio Agamben's important essay *The End of the Poem* argues that poems are "grounded in the perception of the limits and endings," limits and endings which they "define—without ever fully coinciding with"; and secondly, that this problem of endings and limits is an "intermittent dispute with sonorous (or graphic) units and semantic units."[11] Such a dispute returns again to the problematic of the undecidable addressing of poetic form: the poem speaks but to whom is it addressed? Agamben concludes that the end of the poem as "the ultimate formal structure perceptible in a poetic text" is a site of linguistic "intensity." Such intensity is felt and in question in the emails I receive from spammers.

The closure and extent of the text is a fundamental poetic problem. The persistence of this problem situates net writing in the indeterminate space of the impossible closure of the book. Digital writing projects such as Alan Sondheim's "The Internet Text" or Kenji Siratori's diverse work pose the problem of formal closure against serial production. These are works entering the space set out in Pierre Guyotat's work, or differently in Robert Duncan's "Passages" where "phrases have both their own meaning and yet belong to the unfolding revelation of a Sentence beyond the work."[12]

The impossible closure of the book is a confused, paradoxical formula that can be restated as Internet or literature. Meaning what? What is meant by "Internet or literature?" Is it an opposition? One or the other? Or is the "or" meant to say that one is the name of the other, that literature is the alternative name, the other name for Internet? Or what?

Well, why digital writing?

Because the "source" or substructures of the digital are like writing. The codes and flip-flops are differential graphematic systems.

No, that's not adequate. The code is unreadable and unread, the flip-flop is beyond human sense, its differences so small and fast and distant that it is at best a figure of writing.

That is the point: the persistence of the figure of writing. The persistence of an ambiguity between data and instruction, the persistence of a basic undecidability in the digital, the persistence of these codes and sources as what is read but remains unreadable. This persistence is writing.

You mean a literary writing. The contents of the net are writing. The tags of Web 2.0, of the great promises of the social media net, are writing. "Word, words, words," as Peter Morville puts it.[13]

The status of writing on the net cannot be separated from the net itself. More precisely, it cannot be separated from the net as a thing and as an event. As a thing that is present and that arrives. The net is composed of messages—it is a composition—and composed for the distribution of messages. Does the net arrive? Does this question mean, what is the net? Where is it? How do we read this question and what do we read to understand it? That is, what is the message of the net? Does the message arrive? And as an event, the net takes place. It is here. Where? And where is it going?

This is a question of poetics—of the production of writing—and of literature, that is, of the becoming-literary of positivities within discursive and material domains, or better: a question of emergent language practices. Maurice Blanchot asked, "Is man capable of a radical interrogation, that is to say, finally, is man *capable* of literature, if literature turns aside and toward the absence of the book?"[14] I ask in turn, is net writing just this capability?

When "Mrs. Elizabeth Etters, a devoted Christian," writes me, she begins, "I have a foundation/Estate uncompleted {what millions of dollars} and need somebody to help me finish it because of my health, everything is available." That is such an odd sentence, like every other spam sentence, an oddness that rings with the poetic, with the ambiguity of language play and the undecidability of fuzzy communication zones. What does it mean that the estate is "uncompleted?" What does the phrase "what million dollars" mean and why is it in curly braces? How about that final "everything is available?" Does not this conclusion suggest the relativization of any reading and reception of this text? There is an asemantic filature across the message. Is this a refusal to speak the truth? Or an insistence of the true state of the text behind the purported truth? It is as if the surface of the message were distorted by the fraudulent intention. We know some of these textual distortions are inserted to bypass Bayesian filters and the like—as I have discussed above—but one way or another, the result is a decay of language to expose writing as nothing but address. On the one hand, this means an emptying of the form of address; on the other hand, it means the continued functioning of address as a form only, as nothing but formalization. As Galloway and Thacker put it in *The Exploit*, "Spam signifies nothing and yet is pure signification."[15]

We can pick at the twisting of the English language that breaks the surface of the spam message. Mrs Mellisa Lewis was diagnosed "for" cancer: no one says "for" cancer. This is clearly a fabulation, a fake, written by someone in Nigeria or one of those places where these emails originate. I am sure, I know that Mrs Mellissa Lewis isn't even her name! We all access this "common knowledge." What is this certainty, this knowledge? How am I sure of the untruth of this text? Is it enough to point to and repeat the epistemic knowledge of spam practices? How can such knowledge be permanent and stable? This knowledge floats ambiently over all net writing, a finely grained epistemic filter that renders every sign both utterly persuasive and utterly fictional. To speak of modes of address as conventions assumes that we decode their history, operations, and formalisms. We are ever so comfortable operating in discursive fields. Discourse means touching the other through the mark, through the fixation of the word. The limit of this decoding is the sheer fact that "address" remains. I am left with address as the trace of the other. Address transcends decoding. Or rather, in the decoding of address we experience and handle the experience of this transcendence through the act of decoding.

Jacques Derrida wrote of the postcard: "its lack or excess of address prepares it to fall into all hands," adding that "the secret appears, but indecipherably."[16] The secret is traced in every net writing. Arrival and reception means occurrence, something taking place in the email that lets us talk of opening or checking our email. A taking place that "reading" only covers over and does not explain.

Is the answer an analysis of the "media technical" features of addressing? Is the concept of an address something internal to an information system, and not to be confused with human addressees? Can information theory—which deals rationally with the positions and flows involved in sender and addressee—contain and account for addressing in net writing? An information source "produces a message or sequence of messages to be communicated to the receiving terminal," writes Claude Shannon.[17] As we know, the message is encoded by a transmitter for a channel, where it is received and decoded for the "destination," that is, the "person (or thing) for whom the message is intended." Shannon deals with the statistical structure of each of these positions given variably noisy conditions. The positions are firm and whole, ready for inhabitation.

End-to-end

The IETF RFC 644 of July 1974, asks the question

How can the recipient of a network mail message be "certain" that the signature (e.g., the name in the "FROM" field) is authentic; that is, that the message is really from whom it claims to be?[18]

The answer offered in this RFC is built on the end-to-end principle that underlies the net. Though the content of email may be dealt with at "higher" levels of the Internet, sending messages is the basic operation of the net. All features of the net are dedicated to ensuring that messages move from point to point. The content and even integrity of the message is secondary to end-to-end transmission. The physical locations of the end points are sockets that define a communication path. RFC 644 proposes tying certainty of signature to the hardware sockets.

> We propose that the receiving process consider the sending process to be a properly authorized (by the sending host) sender of mail only if the sending end of the communication path is (one of) the socket(s) reserved for transmission of authenticated mail.[19]

Of course, this means that any message transmitted from a given socket is marked as an "authentic" message from that socket. The mathematical specification of sender and receiver, of information source and destination, does not resolve the alterity that I encounter in an email message. IETF RFC 2635 addresses spam, arguing that a "culture" of research and education communities dominated the Internet before it was opened to commercialization, determining netiquette and net communication, and that this culture is "deeply embedded in the protocols the network used." It was a culture of exchange and display, of openness to and of messages. This consensual, cultural domain frames all address in net writing.

The more general concept of a computer address is no help either. An address is a specific binary number that identifies a location in memory from which and into which a computer can store and retrieve information. Often an address is a single byte of available memory. In this definition, a computer address at first appears quite material. In this sense, *address space* is a range of discrete numbers as unambiguous and unique identifiers of information. On the CPU, "addressing" deals with various modes for correlating machine language instructions to particular registers. Registers are specific locations on a hardware device storing differentials set by voltage or other material levels according to a digital logic. A compiler must know the correct addressing mode for the compiled code it produces. Writing directly in assembly language requires knowing the correct addressing mode. Addressing deals with the basic instruction set of the processor. Here we are in solid hardware-based media analysis a la Friedrich Kittler, one that promises to ground theoretical questions in the materiality of bytes and flip-flops.

Yet the site of the address is already double and ghostly. Addresses may be data or instructions, may be a number or a command. The mode of addressing is undetermined. As a result, the displacement of "indirect addressing" is basic to computer architecture. The content of the address specified is always potentially a pointer to another address rather than data. Even more complicated is the virtual addressing common to contemporary computers, where virtual memory uses a series of page tables to map physical memory to different addresses. Specific addresses can be swapped and fragmented, split and displaced. Address space is doubled, multiplied, so that virtual memory enables a large range of references and locations, providing the computer a vast space of representation and computation.

There are no limits to these models, no breaks in the symbolic orders that contain us, but this also means no limits to the flows and emissions of dealings with absence of bodies. All email is mine, is for me, and is addressed to me. I am still that "useless passion" described by Sartre.[20] The last word on this is Alan Sondheim, who himself writes in regard to the message and its attendant protocols: "Subjectivity appears precisely in the absence of its call."[21] I pour myself into writing the absence of the net. Is this not a narrative? Is this not a figural relation prior to the requirement and positing of a communications network of senders, receivers, and channels? The undecidability of a chain of discursive elements is a story of encountering the other in the net's writing, a story of being addressed. Again, what is address? To even begin to answer, to start on that journey, one finds address that cannot be separated from a narrative, from an autobiographical story of the subject. Every address involves a momentary displacement toward the other. Not a spacing or Heideggerian *ereignis* but a tempo of announcement nonetheless.

The existence of the net is inseparable from the third party. There are two stories of this third party: as a protocological institution and also as the absolutely other in the great beyond the net never touches. Every address already comes to and goes from the third party. Not you and I, who already speak and have voices, but that other who listens and to whom our voices drift and carry. It may sound like surveillance, like the fact that the NSA and Homeland Security are always watching, that Echelon—that mysterious monitory network that reads all our emails and exchanges—is at work. Yet surveillance of this sort is a code that can be cracked. To say that I am on the net only if I am viewed by the administrative other of Homeland Security is in fact to know quite certainly where I am. No matter how difficult it may be to determine who is reading and archiving my emails, it is in principle possible. There is a reader and an archive out there.

The problem of the third person is not of this sort. Nor is this the problem of a masquerade where each of us becomes an anonymous third person on the net. It is true that the "I" I assume on the net is very much a fictional "I," that no one knows I am a dog, and I can be anything. I take an avatar and re-make myself. In this instance, as well, it is clear that I remain outside the net, while my masquerade is enabled by the modulation and play of various codes of identities. The other on the net, in this sense, always remains a façade. In the case of surveillance, the administrative other is always watching. In the case of masquerade, only a fictional other is watching.

All this is true—surveillance, masquerade—only in terms of the fictional capabilities of address. To know that I am watched or to know that the other is a façade is to posit an epistemology and to produce a discourse on the fictionality of net addressing. These positions—surveillance, masquerade—preserve the reality of the other beyond the net. Address locates. It opens. It closes at the same time. Address presents and withdraws. It is indifferent.

The indifference of the message can be explained as a flat technical surface, but every surface and every opacity is a psychic field, a folding of the self's interiority. All this affirms the operation and space of the book as the absence of the book, the book beyond the book, in Edmond Jabès' sense. Is this not the point of Ted Nelson when he invents the notion of the link on the model of literature ("literature is debugged" he writes)?[22] Is it not the indifference of literature, or its exterior positivity, that makes it the disruptive paradigm of net writing (and casts it against all Nelson's subsequent efforts)? Literature is an institutional space that creates both an affect of address and a disappearance of address. Literature must be a problem or else it is not worth creating and studying. There is a study of genres of net literature, histories, and practices, which I leave to its scholars. Is it not possible to study the works of literature as a canon of belles-lettres, on the one hand, and to understand literature's emergence, on the other? Is this emergence not what ultimately draws us to the net?

Just as much as it is easy to be cynical about spam email, it is easy to be good and righteous as well. Today Judith Alexander writes, "Help me carry out my last wish. With your help; I want to donate to the needy, the poor and motherless baby's homes. Reply if you can help." I want to give. Does her writing touch to me? Do I read here an expression of the other's suffering and in this a trace of the body? Is this reading that collapses into the immediate sensation of the other's pain?

Perhaps I should give to Judith Alexander. Is not my righteous goodwill also a cynical position? Is this not the cynicism and calculation of reason? Not simply that I calculated the amount to give and made my giving a matter of calculation, but because I responded to the highly codified summons of the text message, the email solicitation, the public radio call.

PLEASE REPLY MY BELOVED. The phrase is emptied and voided. I can discard it. It is easy to be righteous and good, just as easy as it is to be cynical. I am cynical and distanced in refusing to give to Mrs Mellisa Lewis. Who would deny that both give me a control of relations, a position from and in which to place myself? I am comfortable and able to control the distances between myself and the spammer. There is a purity and immediacy with the spammer as well: I know with certainty how to read their message. I know how to decode the salutation and the heartbreak. I do not care about this fake cancer and death. Were they not suffering before?

Michael Berry's *Greetings in the Name of Jesus: The Scambaiter Letters* is a horrific book and a classic of Internet literature.[23] In its complex play of identities negotiated and exchanged on the web, it is a pure example of literature that the net makes possible. It chronicles Berry's turning the tables on the spammers. The reversals are firmly situated within the symbolic positions of the communications circuits. The senders become receivers. They assume the position. Berry's villain's gallery of spammer photos emphasizes humiliation and abjection. Bodies are splayed for our laughter and scorn. We know, with certainty, that these are criminals. We know more than they do.

The condition of the text—the condition of all texts—cannot be separated from the condition of bodies. Learn this if nothing else. I deploy terminology of psyche and body because of the recognition of interior processing that occurs in every address—which is to say, in every mark—a recognition keyed to interior churning and working over the other's body. I get off on their injuries, their death, and their crushed bodies. The sight of it in the news puts me in my place. Soon, the very thought of their disaster, all the disasters, confirms me in my goodness and caring. Look, don't put this on me: you get off on it too. We take our pleasure in being constituted, positioned, and sited in relation to the other.

To do so—to take this pleasure, as you and I both do—is to deal with a bloody, pulped text. The pulped text is one we take in and incorporate. We work through and understand, but more than this, we consume the body of the other. In consuming we build and armor our own bodies. We site ourselves in the communications circuitry.

Technical definitions and RFC protocols and the like are codifications of the consensual encounter with the other. Call this alterity the neutral of the net. If I were not so serious, perhaps I would write, the ne(u)t(ral). The neutral is an uncertain openness to the stranger, to the other as stranger. The fact of the net and my knowledge of it is an impossible neutral persistence of the other, affirmed despite and beyond the codifications of protocols and interfaces.

To send, to address, and to transmit, to receive: all this is palpitation of the other's body. To caress and fold and follow the lines of the surfaces of that other body. The line is the figure of my imagining your body. I am always following this line toward you in the pure unmediation of the flesh.

VII

Logging in and getting off

*Let me cite: in a posthumous fragment,
Sigmund Freud writes: "Psyche is extended;
knows nothing about it."[1]*

Let me state: I write away, as in wear away, one drifting piece of our inconceivable inhabitation of the net. Writing away through myself and toward myself and in lieu of myself. Writing as weapon and spew and nothing.

Let me ask: how many times a day do you do it? Once, twice, seven, eighteen, fifty…? Are your hands tired, your fingers calloused, are you out of breath? We all do it. It is, at the least, a shared experience, more or less habitual, our fingers finding the way automatically. What is it? I am writing on the (k)not of logging the subject of the net.

Let me admit: we all do it, and yet it is hard to describe. What is logon like? It is difficult to find analogies or remediations from other media forms. For example, if logon seems similar to the signature, to the inscription and circulation of proper names, say in publication or in an accounting book, it differs because I inhabit logon. I am told, "You must be logged in to do that." I not only represent myself online but act through logon. Is "represent" even adequate here? Can we think logon in terms of a singular and situated subject who logs on to the network, the subject who is "over here" and who logs on to the network "over there?" Logon is not simple. Is it not a science fiction scenario of being elsewhere, loaded onto the computer? We must acknowledge its fundamental oddness. You may set the machine to log in automatically, say on powering the system up, and you may set up to log out automatically. You may also physically log out, by turning off the machine, which differs from the protocol of logging out of an account. You can also be logged on but not there. You can log in and go away for coffee or go to sleep. You may also log on with several accounts, split yourself across the net.

It is a question of who or what logs on, or is logged on. Am I logged on? Yes, I am logged on. No, it logs on, it is logged on.

Let me propose three formulas about logon. Number three, as follows: log in me more than me. Or as Freud quotes a patient, "It shot through me [...] There was something in me at that moment that was stronger than me."[2] To log on is to mark your occupation of an otherness that always already haunts your imaginary.

To arrive at the second formula on logon, I differentiate log on (two words) as the action or verb, from logon (one word) as the state or the modality of being online. The action requires a story of logon, a narrative and an accounting—and it comes down to accounting—of the labor of logon. Logon is necessary because of accountancy. You log on and begin spending resources. The differential between logon and -off is a condition of capitalization. The meter is running as you surf the net, even if you do not notice it. With this in mind, the formula echoes *Neuromancer*-style: "logon as jack in and jack off." The labor of logon is a matter of putting to work excessive affect and surplus pleasure, of mobilizing body parts on the interface, of the productivity of organs and fluids that are donated to but never delivered to the net.

Still ahead of myself. Formula one: "logon is part of writing my self into the net." This first formulaic statement makes logon a project of being and presence on the net, of construal and working through the net's otherness, where the net is "always on," as defined in RFC 1, the very first of the Internet "Requests for Communication" that set out the protocols and administrations of the Internet.[3] All that the net nets is smoothed and hardened and purified. You log in to a UNIX system and the shell executes files called .login and .cshrc, or something similar depending on the version of UNIX. These files contain your profile, defining your permissions to access other parts of the system, your home directory, your editor and printer, and so on. Logon authenticates you to the system and supplies your username with necessary levels of credentials to interact with the system. Login is tied to permission, to a protocol that recognizes and places a proper name. I present my papers, own up to my name, and ask for my identity to be validated.

Login is productive. What does this mean? First, login is produced, it is a performance framed by institutional and technical parameters. I log in as a persona or avatar. The name I assume when I log in is never my own even if it is mine. Login is fictional and generates a narrative knot that unfolds throughout my being online. Login is tied to the problem of electronic literature. Logon is not only produced but it also produces. It requires the presentation of a working body. It produces work, digital work, work characterized by the problem of the work.

The concept of "logging on" emerges from measuring work. Most histories refer to "clocking in," the procedure of tracking hours worked by punching the clock. A worker arriving at a factory used a timecard to track the beginning and end of work. The card was put in a slot and stamped with the official time. The Bundy Manufacturing Company, the inventor and earliest manufacturer of time clocks, was later incorporated into the Computing Tabulating Recording Company, which in turn became International Business Machines or IBM in 1924. IBM's website still maintains archival documents on setting time clocks.[4] The scarce resources of early computer systems implemented logon as a similar accounting for time on and off the system. Logon as creation of a stock of labor and as measured accountancy of net time is never far away. As both laboring production of the self and excessive display of otherness, logon is captured and canalized in a (k)notting or milieu or apparatus of CAPTCHA.

CAPTCHA

You are familiar with CAPTCHA even if you do not recognize the name. They are the words, phrases, or occasionally numbers, which you are required to recognize and type into forms on websites in order to gain permission for an account or to post to the site. CAPTCHA are now everywhere on the net. Examples of words and phrases, collected over the last few days: lacking Katarina, animal Brothers, Bellerman Dunn, Smillie ury, St Chapin, season Bambrick, m Denounces, and qphwvhag bivkjdi dpokf. CAPTCHA is a program that runs every time the page is loaded into the browser and displays new combinations of letters or words chosen according to a procedure. It selects a word from a dictionary or generates a pseudo-word. It then uses Gimpy or some other program to distort the fonts and other visual features, with the goal of preventing computerized image recognition. The letters are twisted and warped, crossed with visual noise such as dots or hatching, and washed with wildly varying colors.

There are different genres of CAPTCHA. For example, Facebook tends to have phrases, Google curious words, and so on. These are determined by the stochastic process used. Remember that "stochastic" refers to a process that unfolds non-deterministically, where the relation between a moment and a series, or between inputs and outputs, appears random, complex, and multiply determined. Some CAPTCHA programs stochastically produce strings of letters that do not approximate any English language series of symbols. They do not resemble words at all. Others produce a series where letters and digraphs (or letter pairs) approximate English language, so that the string appears word-like. Others produce phrases in a similar way, while some CAPTCHA programs produce phrases from words existing in the language but paired in a stochastic manner. The result is an unexpected but not impossible phrase.

CAPTCHA is an enormous distributed writing machine. It inscribes writing to be read and re-written in a constant and incalculable flow across the net. Where else is there such a repeated engagement with a language-oriented avant-garde non-semantic text production? CAPTCHA relates in a direct but problematic way to contemporary writing practices and to practices of electronic literature. This is especially so as the computer increasingly provides writers new fields of visual and stochastic text production. The "lower-order" CAPTCHA, where stochastic processes offer dense strings unrecognizable as English words, appears similar to sound poetry or concrete poetry. This visual poetry or VISPO connection is reinforced by the combination of words with distortion, positioning in space, color, and non-linguistic visual elements. Another direction, articulated with concrete poetry, is the strain of visual and conceptual art working with scripts at the edge of readability and recognition. Henry Flynt's "The Counting Stands," for example, is a series of artworks that force awareness and difficult engagement with the cultural technology of recognizing and working with numbers.[5] In turn, the "higher-order" CAPTCHA that utilizes phrases and possibly even regular English words approximates L=A=N=G=U=A=G=E poetry and other strains of avant-garde contemporary poetics, with its semantic play and asyntactic construction. Take a look at the dense pseudo-language of David Melnick's *Pcoet* and then read Bruce Andrews' *I Don't Have Any Paper So Shut Up*, with its humorous but seemingly found phrases such as "cheerful robots" or "garage guilt."[6] Both read as collections of CAPTCHA tests. Close the loop by comparing to Patrick Swieskowski's program that scrapes CAPTCHA imagery from AOL instant messenger sign-up screens and presents them as concrete poetry.[7]

Aram Bartoll of the group Free Art and Technology created a series of custom-made business cards of CAPTCHA images, clearly playing on the signature-effect or recognition of the name of the subject accompanying CAPTCHA.[8] He also spraypaints CAPTCHA images onto foam cutouts and pasted them in urban settings. Becky Stern collects CAPCHA images and paints them in acrylic on 5″x7″ canvases and sells them.[9] Gus23's "Dreamcaptcha's combine the magical, protective function of the Native American dreamcatcher with the CAPTCHA image's function of protecting and maintaining the subject's identity."[10]

What is the purpose of CAPTCHA? CAPTCHA is informed by a basic cryptographic principle: it proves you are who you are by making a challenge that forces you to manifest a response that only you can have. You are hailed and called. You must respond, and not with just any response but a specific one. In the case of cryptography, the response is the key that you employ to decrypt the message, but in the case of CAPTCHA it is some task that humans can perform regularly and machines cannot. One possibility is to pose challenging questions, but questions such as "what is the date?" or mathematical puzzles are not easily generated automatically and can be cracked by AI. The ideal questions are not cognitive, that is, they do not involve looking up information in your memory. We know computers look up information faster and with far greater resources than humans. Any computer on the net can randomly generate CAPTCHA responses with the hope of a brute force solution. Even more, one of the paradoxes of CAPTCHA is that the computer running it must know the answer to the puzzle, since it must be able to look up the answer in order to evaluate the test. In short, CAPTCHA generates and grades a test that it itself cannot pass. Inversely and paradoxically, humans can take the test and not pass but always make the grade; that is, we continue to be human even if we do not pass. What would it mean for us to fail the test? Can we fail to be human? What unsurmountable and unfailing quality is always there?

The testing never stops. We cannot fail yet we are never allowed to finally pass. The test is everywhere. Think of spam with an embedded image, perhaps trying to sell Viagra. The spam image is not caught by the email filter, which can only recognize text-based spam. You see it and recognize it. Once again, you are faced with a CAPTCHA test. You may or may not respond, but it calls you one way or another. The spammer relies on the fact that humans perceive to read, whereas the email filter only parses patterns of texts. The same is true of any image on the net.

Of course, CAPTCHA is always susceptible to being broken by programs for image processing. The original techniques for distorting CAPTCHA images were examples of typical problems in image processing and optical character recognition. Sophisticated new programs can find a way to sort through the visual distortions and recognize the image. The result is a kind of "arms race" where spammers and the rest continually develop new programs better able to beat CAPTCHAs. Such a program beats the test by gridding the image, clustering elements, and solving each cluster. It breaks the visual into a raster through a process of "segmentation," and alters the grain of the raster until there is a recognizable solution. It looks for contours, colors, adjacencies, edges, and textures. It is not a matter of recognition of wholes but of graphable regions. It medicalizes the visual into clusters of symptoms.

By contrast to this grid and process, the very concept of CAPTCHA assumes recognition of significant images as proper to the human. The assumption ties CAPTCHA to a phenomenal practice of certain kind of being. It assumes some interiority and responsibility within perception. At the same time, the event of recognition as perception and responsibility is elusive and ever-receding. It is this fuzzy overlap of iterability and signature that articulates the problem of the literary on the net. The affinities to literary production do not let us qualify CAPTCHA as literary writing in a vague sense that everything is potential literature, though this may be true, but point to sites of "the becoming-literary of the literal" in the commerce between human activity and machine technics.

Recognition of CAPTCHA operates with a commonsense concept of the equivalence of an external object with an idea already contained in the mind of a singular individual. Such a concept is the outcome and image of recognition as production. In what follows, I situate this inventive production in flows of intensities that I call display. In the background are theories of semantic morphology, zoosemiotics, and somatolysis, following work by Thomas Sebeok, Alfred Portmann, and others, for whom the surface is not a border or barrier to the interior but "an organ with entirely new potentialities."[11]

Telling humans and computers apart feeds several channelings of desire into and on the net. First desire: to ensure that humans are nodes sited in the net, nodes that can make promises and so become signs producing work. The filtering of humans from programs, with the goal of ensuring that only an authenticated human is logged in, also maintains a model of the consumer. It construes the consumer as a site or unitary subject with a finite repertoire of desires and tastes, codified in the user profile or login information. This construal is tied to CAPTCHA's initial appearance. The Alta Vista search engine began using it around 1997 or so, claiming it helped prevent spambots from automatically submitting URLs. CAPTCHA also appeared at about the same time in Yahoo chat rooms, where more and more bots were entering into the chat. Both problems are commercial in origin. The bots in chat rooms are intended to bring people to consumer sites, just as the URLs are in Alta Vista. CAPTCHA is not just differentiation of human and machine but also differentiation of consumer and marketer. The creators of web-based ads, still a primary online business model, require humans to see their ads and desire their products and respond through purchasing. CAPTCHA controls desiring flows and regulates accelerated incentivization of eyeballs and click-throughs.

CAPTCHA also channels desire to create communality. The networking claims of Web 2.0 require investment and circulation of signifiers of human sociality, ranging from Facebook updates to Twitter tweets on what you are doing right now, all in order to create relations between users. The direction and outcome of these relations are various. They may be leveraged for marketing just as in the Web 1.0 model, or they may be in the service of more utopian goals. In either case, the spambot postings that CAPTCHA seeks to prevent arguably do not contribute to such communality, whereas human eyes on Web 2.0 content creates loyalty to the site and participation in the social network.

Computer and information scientists researching stochastic text production refer to CAPTCHA and the spam it is intended to prevent as "inauthentic text," that is, writing produced by a computer that appears to be written by a human. The research on inauthentic text works with a notion of a simulacrum of surface meaning, or an appearance of authenticity, built on an underlying lack of authenticity and meaninglessness. Inauthentic text is "syntactically correct sentences such that the text as a whole is not meaningful," according to the Proceedings of the Sixth SIAM International Conference on Data Mining.[12] Take a look at the research done at Indiana University, for example, focused on statistical measures for detecting authenticity in text. A related example is work done with anti-plagiarism software such as "TurnItIn," with its measurements of "originality." In all these cases, you find a taken-for-granted relation between the human, on the one hand, and a necessary articulation of surface meaning and depth, on the other. The notion is that humans produce texts that are meaningful no matter how apparently nonsensical they appear on the surface. This is authenticity. The result is a collapsing of the authentication of a signature—text as signed by an author or authority as authentic—with the semantic value of a text, where the signature may be occluded or not present at all. The Indiana University researchers stay on the surface. They fail to recognize that, in CAPTCHA, it is not the semantics of the text that authenticates but the rewriting and repetition, although there are ways that this iteration does frame the text semantically. From emptiness to authentication through rewriting, it is this repetition, this efficacy, this project of a series of excessive displays, that makes CAPTCHA part of logon.

The basic claim of CAPTCHA is captured in the name that forms the acronym "Completely Automated Public Turing Test to Tell Computers and Humans Apart." The translation of the Turing Test into an online challenge and response system to deal with spam occurs in "Verification of a Human in the Loop or Identification via a Turing

Test," a paper by Moni Naor of the Weizmann Institute of Science. The CAPTCHA acronym is developed later; but what is in this name?

For starters, what is a Turing Test? It was originally described in Alan Turing's 1950 article "Computing Machinery and Human Intelligence." We all know the story: Turing is credited with designing a computational machine that broke the German Enigma codes during the Second World War, and subsequently with the development of the Mark 1, one of the earliest computers. Let me describe a version of the Turing Test with more recent media. You engage in an exchange of emails or instant messages or Facebook updates. Is the unknown respondent another person or is it a bot? Is it someone or is it a computer programming passing as a person? You want to know. Based only on the conversation, can you judge whether the other is human or machine? Is there something in what is said or how it is said that differentiates people from programs? If this is a version of the Turing Test, then passing the test is when you are consistently unable to tell whether the other is a person or machine. For the other to pass the test, whether machine or person, is recognition of intelligence. Despite other attempts at definition, the Turing Test is still the single standard for computing machinery and human intelligence.

CAPTCHA is a Turing Test for a human filling out the form. It is designed to fail machines attempting the test. Note that while Turing casts the test as a question of intelligence—remember the essay title is "Computing Machinery and Human Intelligence"— the test is really not an epistemological question of whether a machine can know and how much, but an ontological question of what kind of intelligence a machine can have and whether it is the same kind humans have, where intelligence becomes a measure of existence. Once again, the computer knows the answer to the test— it must in order to evaluate the response—but it cannot pass the test. Turing ties the existence question to a good enough imitation of the human. Imitation of what? For Turing, humans too must create a good enough imitation of being human to convince others that they are human. In fact, what distinguishes being human is this imitation and not the fleshy body producing it. Perhaps this is not so paradoxical, since it follows the same strange loop of reentrant logic employed by René Descartes, where "I think therefore I am" means that the Cartesian thing that thinks, thinks itself thinking, and therefore thinks, and so on.[13]

Keep in mind that this problematic of imitation, dissimulation, and masquerade is predicated on the symbolics of the name. The imitative auto-production of the human in the Turing Test can be seen in variations on CAPTCHA. The CAPTCHA acronym was coined by Carnegie Mellon University computer scientists, based on Naor's proposal. The key figures here are Manuel Blum and his student Louis von Ahn, who patented the term "CAPTCHA" and sought to leverage the test both to solve a variety of computer science problems and to produce work. Adapting the concept of work expended through a computer's microprocessor, von Ahn sees our standard interactions with the net as "wasted cycles."[14] We waste time playing *Minesweeper* or *World of Warcraft* or surfing the net or downloading music. For von Ahn, we are "parasites" on the net and should instead be in a symbiotic relation with the machine. The language of parasitology evokes *The Matrix*, where machines also evaluate humans as parasites, in order to make use of us as batteries or power-generators. Von Ahn's cyborg position is similar. He seeks to convert the excessiveness of human display on the net from a series of intensities to sites of recognition and desire, and through this conversion to create productive and commodifiable flows, and ultimately to create an abstract stock of net labor through logon.

One account estimates that "existing CAPTCHA systems represent approximately 150,000 hours of labor per day that could be transparently tapped into. [...] That's approximately 75 years of normal, full-time work accomplished every day."[15] Consider the time spent recognizing and typing and solving CAPTCHAs on the net. Why not make this productive labor and put CAPTCHA to work? Consider book digitization and scanning. Project Gutenberg, the oldest attempt to digitize full texts of public domain books, adds approximately fifty new works per week, and Google Books reports scanning books at 1000 pages per hour. At the same time, optical character recognition, or OCR, is imperfect and the digital image is never the same as the text scanned. Who can keep up with the proofreading? Von Ahn asks, how about CAPTCHA as distributed proofreading to resolve the bottleneck? Take a word that OCR software indicates is undetectable or in question and present it in a CAPTCHA. The word can be alone in the CAPTCHA image or there could be a combination of random strings and words to be proofed. The human identifying and re-typing the decontextualized word is proofreading. If several people type the same response to the word, it can be considered proofed. It is authenticated and the computer can store the value of the word in the OCR text. The process requires linking the CAPTCHA program to particular locations via the net to make sure that work is coordinated, so that the word is supplied and the answer returned back to the proofreading location, and so on.

This conversion of the intensity of display into the institutional codification of desire, and in turn into flows of productive labor, is underlined by the purported use of the method in websites with pornographic images.[16] In this case, a word needs proofreading and is copied to the verification page of the porn site, with the notion that people wanting to access porn are willing type in a few measly words. The desire for the porn images smears into the CAPTCHA images and propels the proofreading.

A more systematic approach is von Ahn's "reCAPTCHA" system patented at Carnegie Mellon. The reCAPTCHA results are assembled as an e-book and submitted to Project Guttenberg, automating existing distributed proofreading efforts, which claim 14,415 texts proofed as of February 3, 2009, the first being an 1884 "Aesop's Fables" completed in July 2006. Facebook, Craigslist, StumbleOn, and other popular sites all use reCAPTCHA.[17]

More recently, von Ahn began working on the related problem of labeling images with words. Most current image searching on the net looks for file names or requires humans to create labels. There is no reliable computer program that will process an image, and label it with tags and other descriptions. Rather than manually create labels for images, von Ahn created a game called "gwap" that employs CAPTCHA-like image recognitions and uses the results as labels. You and a partner see an image and are asked to label it. If you both give the same labels, you win the game. Following a certain number of wins on a given image, the labels are stored as *verified* and *correct*. Verification means a shared encounter and recognition within the specified domain of the image.

The approach adapts Web 2.0 crowd-sourcing and voting systems, familiar from applications such as Flickr, which is essentially a vast system for labeling photos, and also techniques from applications such as "hot or not" and "rank your professor." Von Ahn claims 75,000 players so far, who created 15 million labels. He describes people playing the game over 20 hours per week, even up to 15 hours straight.[18] At this rate, 5000 people playing simultaneously could label all the images in Google and soon every image on the web.

Or consider Amazon Web Services' "Mechanical Turk," billed as a "marketplace for work that requires human intelligence." The workers or "turkers" perform "human intelligence tasks" (HITs). As the website puts it:

> There are still many things that human beings can do much more effectively than computers, such as identifying objects in a photo or video, performing data de-duplication, transcribing audio recordings or researching data details.[19]

Anyone can sign up as a turker and access the many tasks that Amazon brokers for companies, who would traditionally have accomplished these tasks by hiring a large temporary workforce, which Amazon points out is "time consuming, expensive, and difficult to scale." The result is an "on-demand workforce" to "lower cost structure," which allows laying off traditionally employed workers. Most of the tasks are similar to CAPTCHAs: transcribing texts, labeling images, and so on. A twenty-five-word description of a piece of hardware pays five cents, while finding the image of a box of vitamins on a website pays two cents. The tasks also include searching satellite data for missing persons, notably in the search for the aviator Steve Fossett, but there seems to be little success in upwardly organizing the dispersed network of turkers for such critical and focused tasks. Critics see the Amazon Mechanical Turk as a "virtual sweatshop."[20] The workers are treated as contractors and do not pay taxes or receive minimum wages and other benefits. Further, it is true that most of the HITs are boring, repetitive, insulting to the intelligence, and pay literally pennies. One can work all day and earn less than two dollars, though it is possible to make considerable money.

A number of recent artworks use the Mechanical Turk. Clement Valla re-conceptualizes the instructional aspect of Sol LeWitt's work by paying the turkers to recreate a LeWitt drawing.[21] Fred Benenson recently started a number of art projects using the Mechanical Turk. In *Emoji Dick*, Benenson pays Turker to translate each of *Moby Dick*'s 6438 sentence three times into the Emoji icons used in Japanese cell phones. The public will vote on the best translations, which appear in the completed artwork. Finally,

Aaron Koblin's numerous projects in this area include "The Sheep Market," where workers were paid 2 cents to "draw a sheep facing to the left," and then collected the work and animated it. The later "Ten Thousand Cents," in collaboration with Takashi Kawashima, tasked out tiny drawings to workers, that—when assembled—formed a $100 bill.[22] I am not able to discuss these works at length here, except to note the explicit critique in the notion of the turkers as sheep or as unwitting forgers.

What makes a person be a turker? Perhaps to make money, but despite Amazon's rhetoric of the on-demand workforce, if you take a quick look at Turk Nation, the website for the community of turkers, you find many pursue HITs for the thrill and distraction of solving the problem. In short, they do it to display themselves to the otherness of the net, in the same intensity that one gets in CAPTCHA.

Taking the test

The Turing Test purports to distinguish human and computer, to give you the privilege of asserting your humanity and unique human identity. It offers a Cartesian closure where I discover myself in a circuit through the machine. Jacques Lacan once wrote that if "I press an electric button and a light goes on, there is a response only to my desire."[23] The truth is elsewhere: we traverse an apparatus that distinguishes potentials across a field of intensities, and we produce a discourse naming these potentials as human and machine. Keep in mind that the CAPTCHA method is not limited to words and phrases, but could be used with any symbol, including images. Anything can be the object of the recognition and intensity of a CAPTCHA. CAPTCHA is a writing machine that never stops writing and rewriting everything into the symbolic order.

Turing's 1929 paper "On Computable Numbers" conceived of the computer as just such a writing-reading machine, where paper with written symbols passes before a reading head or eye that notes a symbol, stores it, and operates upon it. The symbols must be discrete in order to be grasped in a glance by the reading head. In turn, any discrete symbol, symbolizing anything that can be symbolized discretely, and any chain of discrete symbols it may be a part of, is computable by such a machine. The game of

twenty years later treats human and machine as similarly discrete
and computable, insofar as both produce discourse as chains of
discrete symbols. A computer is a "discrete-state machine," writes
Turing, a machine which moves "by sudden jumps or clicks from
one quite definite state to another."[24] Discreteness is fundamental
to the computer and to the digital in general. For Turing, machines
and humans can be distinguished as long as they are conveniently
defined as discrete entities.

Turing adds, however, "Strictly speaking there are no such
machines. Everything really moves continuously."[25] Nothing is
discrete. Writing does not employ discrete symbols, although
its outcomes may be read this way. The writing machine is not
a discrete apparatus in Turing's sense. Ink bleeds into paper. The
computer overheats and sheds light and loses data. Computers
exist in a narrowly controlled thermal range. The website for
Intel Corporation, designer of many of the world computer
microprocessors, states that "thermal management" refers to "two
major elements: a heat sink properly mounted to the processor,
and effective airflow through the system chassis. The ultimate goal
of thermal management is to keep the processor at or below its
maximum operating temperature."[26] The computer is not discrete
and differentiated but is a radiating sun in a constant state of decay.
Using a computer is handling the decay and dispersal of thermal
management.

There is no discrete but there are series. Series drift and distribute,
but there is a dreamed and delirious communication between them.
Rather than the simple distinction of computer and machine in the
Turing Test, and its transposition to CAPTCHA, human positions
in the apparatus are heterogeneous and wayward. There are at least
four positions for the subject of the net, and two differential writing
series across these positions.

Certainly, there is the computer that fails the test and does not
recognize the image. There is also the threatening *Blade Runner*-
esque replicant computer that passes the Turing or Voight-Kampf test
and requires ever more sophisticated CAPTCHA programs. There
are humans that pass the test, and this means not only recognizing
the image but also affirming this recognition and answering to the
desire or call of the program. But I may not answer; that is, there
can also be a human that recognizes but does not respond, refuses
to rewrite or writes something else altogether.

I might be human but not recognize the image. This could be a problem of ability. The W3 points out that the ubiquity of CAPTCHA and the reliance on visual images "presents a major problem to users who are blind, have low vision, or have a learning disability such as dyslexia."[27] Suggested alternatives include logic puzzles that would also be unsolvable by computers but still recognizable by disabled users; sound output of the sort already implemented with mixed success at Google and Hotmail, where mousing over a small disabled icon speaks the CAPTCHA text aloud; limited-use accounts, where unverified access would not threaten the system; and non-interactive checks that use system-wide measures to identify spam and other threats rather than preventing these threats at the point of posting or signing up for an account.

The philosophical claim of CAPTCHA is reinforced by the calls for web standards for accessibility. The technologists designing CAPTCHA truly want to make it accessible to all humans without regard to ability. The intention is to separate ability or disability from being human. Anyone human should be able to pass the CAPTCHA. It does remain the case that certain disabilities will still prevent a human from passing the test. Similarly, I might recognize the CAPTCHA but make a mistake in typing. Perhaps my hands are shaking from too much whisky or perhaps my keyboard is malfunctioning. In any case, I can re-take the test. I can call forth another CAPTCHA instance. The CAPTCHA does not aim to exclude. It wants me to manifest my human-ness. The test solicits this display.

In typing the CAPTCHA text into the online form, I meet the other's desire. I give what it wants: not the re-written nonsense text but the confirmation that a human act of recognition occurs. In responding, I signify that I am capable of promising, capable of regulating myself in accordance with a protocol. Being human is tied to this promising.

What is recognition anyway? I can recognize the string and type something else. Why should I respond to CAPTCHA? I can respond with my own text. I can refuse every CAPTCHA I encounter. I exclude myself from the net and it does not welcome me. I exit from its domain and refuse to meet the desire of the other, refuse to give the response it asks, and refuse to be confirmed in my human-ness. I am parallel to the inauthentic position of the replicant or transgressing computer that passes the test through some sophisticated optical recognition software.

Bring to mind the well-established critique of the gender complex at work in Turing's test. Turing does not hide this, but announces from the first that he bases his test on a parlor game where the goal is to distinguish the gender of another only by means of written notes. It is gender difference that Turing's game plays on, a difference already assumed by the time of the test. I must add, gender difference and sexual differentiation as the condition of a body of your own. Consider the latest iteration of von Ahn's game: a "gender guesser," that will guess your gender based on how you label images. Or consider Naor's initial proposals for possible test methods, before he arrived at his recommendation of the CAPTCHA-like text recognition and re-typing approach. All Naor's alternative proposals strikingly iterate a human invested in the performative protocols of the civilized and sexed and gendered body, all of which a computer is presumed not to posses and human necessarily to possess. He proposes tests built around "gender recognition," "facial expression understanding," "finding body parts," "deciding nudity" (that is, deciding if a person in a picture is nude or not), "naïve drawing understanding," "handwriting understanding," "speech recognition," "filling in words," and "disambiguation."[28] Link in Turing's homosexuality, with the resulting institutional reaction, including enforced drug treatments and eventual suicide, and you see a libidinal apparatus that sites and organizes in order to produce.

The difference thought makes

As Jean-François Lyotard pointed out, echoing Bateson, the difference that makes the difference is difference.[29] Human intelligence can go on without a material human body, but not without the differential embodiment of libidinal fields, with folds and drifts that cluster as organs, partial object choices, and sexes. What does the field produce? It produces an utterance that will be its name, its own name. I do not assume a persona in logon. There is no dissimulation of intensities: they are there or are not. It is not that the thinking subject's sexuality is conditioned by differentiation; rather, our differentiation into subjectivities is nothing else other than sexuality. The test writes out this difference. It constantly tests and re-tests singularity by naming or labeling.

The first writing series is circulating names. Logon as a state or modality of being deals with names. An object is an object on the net because it is named or "addressable." This is a fundamental condition of the net. It follows the command shells of UNIX and other operating systems, where the environment is nothing more or less than objects that are manipulated through combinations of names. The simple *ls* command, for example, gives direct and powerful access to the structure of the UNIX environment. To operate in UNIX requires a combination of knowing the proper or true name of an object and possessing the necessary permissions.

Logon is a particular entry into this realm of names. In logon, I leave a trace readable as autobiographical name. This trace is more or less a signifier. If logon is part of the symbolic, it is noticeably split between the username and the password. It signifies in this split. Anyone can have the username. It is an exchangeable token, a given name passed down by sysadmin parents. It is mine because I can give it away. The net gives me the name and confirms me as this name, and for this reason I can give it away, to you or to anyone on the net. The basic symbolic economy of users is constructed on these exchanges. Usernames correspond to accounts on the net, to sites of data. The exchange of names is potential access to these sites, as both the social networking power of Web 2.0 and spam attest.

To hide a secret, such as a password, I can write it down and secrete it physically under a rock or beneath the ground or within my memory or tattooed on my scalp. Nothing is hidden in this way on the net. Everything in this environment is written and exists through writing and in a net of writings. UNIX is a file system. The operating environment is run by files and is itself, as an environment, a representation of the organization and movement through the file system. This means that UNIX is a space of texts, with rules of access and exclusion also determined by texts. The UNIX shell .passwd file is a collection of information about user passwords. Typically, another encrypted shadow file also stores personal information about users. This shadow file is less accessible, perhaps only to sysdmins, but is yet another writing within the net of writings. With every cycle, this text and all the others are re-written by the server and the waves of the net.

In UNIX, a file called the login log stores failed login attempts. In fact, it is only after five unsuccessful attempts that it begins to store attempted logins. The first five attempts are assumed to be mistakes, lapses on the part of the subject who is logging in. After five attempts, the failures are assumed to be intentional and not errors. The system begins to record the attempts in a login log. It writes a file of the attempts, a growing text of ill-formed logins. It records login names, tty specifications (that is, the terminal the attempt is made from), and time of login attempt. The log is related to server log files that provide histories of requests to servers over the net. Every action over the net is recorded, every IP tracked. The webmasters see all that we do, both individually and broadly, in statistical sweeps of activity. The World Wide Web Consortium, or W3, the major organization setting web standards, maintains a standard format for login logs. Such files can be mined for information and leveraged for advertising spam. The login log goes beyond the server log: it records the complete text of my failed login attempts.

Of course, there is the possibility of purposeful failure; of creative failure; of psychotic refusal to log in; of refusal to meet desire of the other; of refusal to present my papers. I instead offer babble and nonsense. I produce a hidden text I will never read but which is the stored kernel of my own otherness vis-a-vis the system. This literary possibility is fundamental to login. There are also parallels with artistic practices intervening in "proper" modes of login; for example, Joseph DeLappe's logging into *America's Army* for his Dead-in-Iraq project, or Annie Abrahams' *Being Human*, with its explorations of online being.

I log on and write a text. The text grows. I add to it daily. It is characterized by repetition and aggregation. It is iterative and spaced, a rhythm occurring in certain moments and hours, for certain times, for certain durations. The logon text, if we could read it, would go like this: username password username password username password, and so on. Or, sbaldwin ＊＊＊＊＊＊＊＊＊＊＊ sbaldwin ＊＊＊＊＊＊＊＊＊＊＊ sbaldwin ＊＊＊＊＊＊＊＊＊＊＊. Already a starred text, a secret and private zone of the password; already a text of depths and inventions, of bodies and codes. The asterisks of the password are crosscut into my body, or are eruptions of something interior onto the surface, like bulging skin visible where I meet the screen. The password is closer to me than the username. You recognize that the password is part of the logon text. It is there, even deeper than the username. It is perhaps the most important part. I can give anyone my username; it is nothing but a severed part object without the password. It takes both to log on, but the password is deeper and closer than the username. The username is discourse, a name given me by others, passed down by parents and sysadmins, and mine because I can give it away. The password is different. An effective password—that is, one both memorable and unguessable—is held within, inaccessible to prying eyes. Or rather it appears as nothing but asterisks that burn with interiority.

All this is played out in a visual field. If you stand at my shoulder and watch, you will see my username but you will not see my password. It is encrypted and asterisked out. Of course, it is possible that you are very quick and able to remember the sequence of my fingers on the keyboard. You can try to occupy my movements, inhabit my fingers as they dance on the keys, but you never grasp the non-visible and hidden field of recognition I share with the machine. It is at most visible as the blot or stain, as erasure of encryption, as the asterisks and markings out of the password. I know and the net knows this non-visible domain of "correctness." I and the net can look through the asterisks to see the password that I give.

To reach this domain requires perversion. For the screen as a place and the computer as a thing, to identify with the image and to channel my pleasure to its recognition, I must perversely give myself to the net's desire. Is not this deviation the very definition of perversion? I modulate my desire and myself as desire in writing to the other's desire.

What difference would it make for me to add my password, to replace the asterisks with the string of characters, to put it into a discourse that is visible and readable? To do so involves exposure. It feels a bit like dropping my drawers. I do not share it with you; it's too intimate. It requires letting you in on the perverse relation to the net. Yet, I've already assumed the position with the system. I am already perversely displayed. The network knows my password, knows it so well that it can encrypt the visible, showing only the string of asterisks. The net not only knows it but is waiting for it, providing a text box for me to insert my password. I meet the net's expectations. We exchange something that only appears as asterisks or stars. The net knows what I want and I know what it wants. It desires nothing but this intimate sharing.

This means that I could tell you my password—and who must you be and what would it take for me to do so?—and you could assume my position in relation to the net but you could never assume the position of the net. You could take my spot and touch the keys and give the net what it wants in my place. You could create the phenomenon of logon in my place, the visibility of the username and the starred text, guaranteeing that in that other place of the net you are on for me, you are me for it.

The scene of logon is obscene. Before any recognition there must be a surface with holes. There is a box, an empty field that I fill in. We are well trained to know where these punctuations and openings are in the screen. We recognize this hole, no longer surface but orifice and entry point. The introjected topology is transcendent beyond the ideology of interface design and interactive elements. The screen must be a depth, a place where I put something. I enter my username and password into the opening, but it cannot be filled. It is hungry. What am I filling in logging on? A hungry hole: is this a mouth or a genital? The hole in the screen must be the rim of an orifice. You are familiar with a range of rimmed orifices. The perversities of the interface make it possibly the opening of anus, ear, nose, vulva; but the orality of logon dominates. You feed the screen, you release the password into the buccal opening that macerates and works over the words. Whose face? There is a crowd here, just as there is at every interface. The face is no one; it is the face of the system. It alone takes in my name and it alone sees my password. The sysadmin can check my password but is not the one who sees it every time I log on. The sysadmin is privileged, it is true, but only for a glimpse at what is always before the gaze of the net. You grow into the screen through the surface of the "skin envelope" described by psychologist Esther Bick.[30] Think of David Cronenberg's *Videodrome*, or of the "extraversion" of Hans Bellmer's dolls.

The content of the other's desire, whether password or CAPTCHA text, is smoothed and void and enigmatic. It is exorbitant simulacrum that communicates with the second writing series: not the exchange of names but body display. A condition of becoming-literary, an intense site of frantic resistance or drive to log on or recognize. What do I offer the computer? I sacrifice my body. I display it. I make my body a sign. I can list the offerings from sacrifice, to display, to the work of code. In recognizing the

CAPTCHA or typing the password, we give our gaze, the weight of the hand on the keyboard or mouse, breathing and movement, the flow of heat and fluids from the body, all expended absolutely at the interface.

You are not used to the machine at this level. You approach it as a social object, as projection machine automatically displaying images and that takes us somewhere else, as in "Where do you want to go today?" "Machine" invokes the order of steam engines and industrial devices; the computer's screen and microprocessors are fundamentally static surfaces, at least to our perception. This automaticity of display and topology of the screen are part of what we mean in calling the computer a machine.

I squeeze the mouse. I give it movements of my hand and arm. What else? Heat of my body; pressure and weight of my hands on the keyboard; my breath, inhalations, and exhalations. Perhaps I offer more. I may carry my laptop with me and hug it close, offering it the curvature of my chest or my leg. I may enter into the erotics of netsex, with fluids across the mouse and desktop. I lick the screen, cut myself and fill the gullies between letters and numbers.

Let me be very clear: as I sit before the computer, all that matters that I can offer it are these gases and fluids. This is all that I can materially communicate. There is no economic exchange at work. I do not exchange these parts for offerings from the computer. There is no circulation. Expulsions, fluids and gases, cast-off body parts, flakes of skin, all toward the computer. All of my offerings are attempted couplings. The couplings are not netsex—this is not yet the net—but they are excesses expelled from my body toward the computer. This relation is below and before the net but a relation to the computer nonetheless. The relation is sacrificial, offered with no response, not a hope. At the level of sacrifice, the computer takes and gives nothing in return. It remains a surface and contour for my offerings. I offer the way my fingers fit to the keyboard. The ergonomics involved code my body offering. My hand grabs the mouse, and this too will be coded in terms of human–computer interaction. The computer takes all that I offer, all that I could possibly offer and more. Sacrifice is pure expenditure. I give and give. All this is pushed forward in the time of my body in the presence of the computer. I offer and shed and present these fluids and residues from the real time of my body.

Pause for a moment and marvel at the expenditure, at the sacrifice given to the screen. The energy, the attention, the devotion. All this is nameless, forgotten, and absorbed.

Is not logon a problem of this fuzzy, messy interface of sacrifice and simulacrum, of communicating series of names and bodies? At the scene of re-writing and iteration there is no recognition but intensity and display and ostentation. Not the confirmation of unique subjectivity and identity but the aftereffect of an apparatus that converts intensity into productivity.

VIII

Plaintext

"Plaintext" is not a clearly defined term. As a starting point, consider the range of definitions in the National Institute of Standards and Technology Glossary of Key Information Security Terms. It starts with the context of cryptography, defining plaintext as "Data input to the Cipher or output from the Inverse Cipher," and secondly as "Intelligible data that has meaning and can be understood without the application of decryption." Thirdly, it also defines it simply as "information."[1] *Internet RFC 2828*, the Internet Security Glossary by R. Shirley, defines plaintext as "Data that is input to and transformed by an encryption process, or that is output by a decryption process."[2]

Note that plaintext is both the source and result and the throughput of the process. Plaintext as part of an encryption process may seem distant from plaintext as just vanilla unformatted text, but the fact is we cannot abstract "text" under conditions of today's information systems from encryption processes. While the IP protocol and the link layer of the net are unsecure, drifting, and open, all higher layers of application and transportation—in short, all layers of human read-write-execution—are encrypted. Indeed, *RFC 2828* self-referentially includes the Internet itself as one entry in the Internet Security Glossary. To use the net is to engage in encryption.

Cryptography, literally "hidden writing," shadows the literary in that it theorizes the partial readability or unreadability of a text passing through transformations and handlings of a writing system. The cryptographic principles for "handling and filing" of plaintext, that is, for "the original, literal texts of messages sent in secret or confidential code," are serious and strict. In his lectures training Signal Intelligence Service and National Security Agency cryptographers, William Friedman states:

> There should be in existence only a definite, limited number of copies of the plain text of the message cryptographed in any secret code or cipher and these copies should be carefully controlled in distribution, handling, and filing. [...] The plain text should never be filed with the cryptogram itself. The further the containers in which the plain-text copies are kept from those in which the cryptograms are kept the better. These containers should never be kept in the same safe. If only one safe is available, it should be used for the plain-text versions, keeping the cryptograms in a locked filling cabinet.

Also:

> Plain text and its equivalent code or cipher text must never appear on the same sheet of paper for final copy or for filing purposes. Work sheets should be destroyed by burning.[3]

What interests me here is the maintenance and labor required to isolate the plaintext from its encryption, a distance that supports technical layers, cultural institutions, but also military logistics and rules of engagement. Such maintenance and labor produces the aesthetics of the text. The force that binds and separates plaintext and encryption requires burning should they brought together. The conflagration built into these texts supports and is supported by law and ideologies of identity and security. The uniqueness of the text is inseparable from this conflagration.

The destiny of plaintext is isolation because of uniqueness, destruction to prevent reading. Plaintext rests on an almost metaphysical topology of materiality and information. It construes an exterior to information systems that interrupts the text with a demand. What demand? Plaintext interrupts to announce a community of writers beyond the military logistics and rules of engagement that determine information systems. The literary is inextricable from this interruption.

Under today's conditions of "Pretty Good Privacy"—where the NSA no doubt intercepts all our communications down to the merest post-it note, but probably cannot be bothered to read the critical mass of data beyond flagging key words such as "turban" and "peace" using data sniffing protocols such as Echelon—under such digital encryption standards, every byte of data—and this means every pixel and every screen of every computer everywhere, that is, every byte and its monitored output—figures in the open secrecy of plaintext and encryption. It is as if plaintext were the secret core of information. It is as if a total decryption would produce a textual superset of all possible knowledge.

Plaintext is the fantasia of the library and the cryptoanalysis of text is a model of the depositing and retrieval of knowledge in archives. Of course, the crucial thing is this "as if" which metaphorically repeats the notion of the archive without clarifying it. There is no plaintext but only systems of encryption and decryption, institutions of secrecy and revelation. Information is always totally secret and totally open.

At the same time, we all know that plaintext is also commonly used as the "lowest common denominator" file format in a system. If I am editing a text with someone else, perhaps using Microsoft Word, I may ask them to send it to me in "plaintext." I expect an unformatted document, written in Wordpad or another text editor. Similarly, plaintext is the default format of email messages. It is a non-format for the simple transmission of a message, a technical format that defines "without format." It is where the message is nothing but itself and is not the medium. It is *amedial*.

This is highly problematic. One implication of this concept of plaintext is inseparability from exemplary proximity to sharers, those to whom we need to get a message, and where format means surplus added to and possibly obscuring the message between friends. My point is that this notion of plaintext requires communities where text is read for the message, where the message is the medium. All of which implies the notion of a minimal communication circuit of nothing but sender and receiver. One thing that characterizes plaintext and ties it to the problem of the literal, and in turn to the problem of the literal qua literary, is this implication and invocation of friends, of a community of lovers of literature, and the implication of this community as a determination of its plain-ness, shifting focus from technicality to community. Even the plainest of text is addressed to another. This means plaintext is never plaintext.

March 11, 1968

President Lyndon B. Johnson approved the American Standard Code for Information Interchange, or ASCII, on March 11, 1968. Johnson was simply institutionalizing what was already the case: from its first implementation in 1963, ASCII was and is the standard for encoding data. All communication, transmission, and storage on computers or other telecommunications devices is ASCII-encoded. All characters must be ASCII-encoded or converted to ASCII and then encoded. Information is the transformation of ASCII encodings. We may think of the web as a zone of diverse multimedia, but its protocols are all coded through and with ASCII. Objects on the net are addressable because of a wrapping of ASCII and it is through ASCII-compliant protocols that communication is possible. It is the medium of our media.

Plaintext as unformatted text typically means ASCII-encoded text. This most basic text is encoded and encrypted, in order for it to appear as text. ASCII is the protocol name for the plaintexts we read. ASCII connotes plaintext and source code, and therefore the ethos of programmers and hackers, in contrast to proprietary formattings and layers used by applications such as Microsoft Word and its.doc filetypes.

At the same time, there is a politics of national characters in ASCII. The initial ASCII lookup table was ordered for English characters, with pounds sterling or kanji as complex exceptions. ASCII is the precedent for the recent Unicode universal character encoding, which assigns its first 128 characters to the ASCII character set, and debates continue over encoding of Korean or Ogham or Klingon characters. The story of escape codes and swap tables for alternate character sets is important but lengthy; my point is the contested and heterogeneous nature of the standard from the first. The problematic relation of universal encoding to local context is captured in the Japanese term "mojibake," which names the incorrect mappings we see as rows of boxes and other glitch character when software attempts to render Japanese script. ASCII and now Unicode provide the framework for the net's phenomenology, determining appearances and conditions of experience.

What does ASCII encode? Perhaps everything. Michel Foucault defined the lower limit of discourse with the example of the given-ness of "a handful of printer's characters."[4] Friedrich Kittler's critique broadened this media *a priori* beyond print technology. ASCII would seem to offer a similar handful of characters, but this is not the case. Instead, it is a heterogeneous field of transformations, a Zermelo formalized set, an immaterial and non-medial a priori. The definition seems simple enough: ASCII is a seven-bit character encoding specifying a relation between characters and bit patterns. Of course, ASCII is an eight bit code but uses seven digital bits to encode each character, from 0000000 to 1111111. I will return to the eighth bit below. ASCII maps a specific character to each value in the entire range, resulting in a total of 128 characters for the binary digits from 0 to 127. The first thirty-two digits (0 to 31) and the final digit (127) are non-printing characters.

ASCII encodings are transformations, mappings of f(a) to f(b), but they are also conditions of appearance. They will always be made phenomenal and material. The non-printing characters memorialize incorporated and optimized bodily skills. One way this is evident is in the need for escape codes or code breaks corresponding to keyboard SHIFT or SPACE keys, accommodating the manipulation and collation of material writing surfaces.

The difference between the DEL and BAK (or backspace) keys is complicated. In a computer where the writing surface is treated as an immaterial support for text, a backspace effectively deletes characters. In the physical medium the backspace pushes the typewriter carriage back a space to allow overtyping or revision. The physical surface is moved incrementally without necessarily affecting the character imprinted there. Does BACKSPACE mean delete or does it mean simply a physical move? Not a straightforward question but a matter of whether you deal with characters floating on an immaterial writing surface, or physical media with character markings.

Many of the non-printing characters are obsolete. Most of the non-printing characters are control codes used to alter physical mechanisms. The Carriage Return character originates with the typewriter, where the bar or button rotates and advances the cylinder or carriage holding the paper, and returns it to the left hand side of the page. In ASCII, the code commands printers to return the cursor to left hand side of the page. The locations, page size, printer type, and so on, are determined by particular output formats. Similarly, the non-printing character Form Feed causes the printer to advance and eject the current page and to begin printing on the next page.

All of these commands alter the physical mechanism that displays ASCII characters. At the same time, the commands themselves remain abstract and separate from particular printers, with the material particularities of manufacturer, parts, wear and tear, and so on.

The remaining ninety-six codes correspond to printable glyphs. These include punctuation marks, symbols, digits, and letters of the alphabet. 1000001 maps to the letter "A," 1000010 to B, and so on. Lower-case and upper-case letters differ by a single bit, so lower-case "a" is 1100001.

The table printed in the original 1963 standard document reinforce the appearance of mapping bits to printed characters. The table's courier typeface, familiar from terminal screens, seems to correspond to the bit patterns, since we are familiar with seeing both displayed in Courier. In truth, this is a matter of the "seeming" of representational practices. In practice, textual output was a secondary goal. The goal of the standard was not to define particular representations but to maximize possible encodings to a broad range of media.

ASCII is an immaterial grid. The 1963 standard did not represent encodings using digits but with a dense grid of eight columns with sixteen characters each, where each row corresponds to a four bit pattern (0000 to 1111) and each column to a three bit pattern (000 to 111). So, the upper left position in the grid was 0000 in the first four bits and 000 in the last four bits, and contained the null character. The bottom right position in the grid was 1111 in the first four bits and 111 in the last four bits, and contained the DEL character. The four columns in the middle of the grid contained the printable characters. As a result, the contents of a column or row could be transcoded into another column or row by simply changing a bit. This was quite intentional and the product of long debate, as some of the standard's appendices make clear, so that symbols commonly used to precede numerical information (such as the #) occupy a separate column from the digits; and so the printable symbols cluster together visibly on the table, preserving the division between the visible and the invisible, the printable and the non-printable, at least in terms of the printed table. The

1967 revision to the ASCII standard added uppercase characters and some additional symbol characters, but the column and the dense centering of the graphic characters were largely preserved. Lowercase and uppercase only differed by a single bit.

As a result, ASCII is less an encoding than a transformation field dictated by the grid, by the printed table, where the position of the code is as important as its content. Codes are grid addresses and can be targeted as such.

The first code in the grid is a null character. The null character is not the number zero, which is code 48 or binary 0110000, nor is it a space character. It does not correspond to the appearance of a blank on the materiality of the screen. But it is also not the absence of inscription on the surface of appearances. Rather, ASCII code 1, digital 0000000, is a code for NOP or no operation. While the space or blank character may have effects in print, the null character may have effects in computer code. In C and other languages, the null character terminates strings. For example, "hello\n" tells the compiler to print hello and then to terminate the operation. The null character corresponds to no printed mark but does correspond to a medial difference. It is a condition of appearance but not a phenomenon in itself. In contrast to the inscription of characters on the material surface, ASCII exists in a dataspace of operations guaranteed by the negativity or syntactic gap of the null character. As Anthony Wilden and Brian Rotman note, true zero is crucial to placing and ordering of discrete digital elements. As the basic act in this space, as the first ASCII code, null mediates all other characters in the grid.

The eighth or parity bit is another digital supplement, a technical specification dictated by the alterity of ASCII. ASCII is often referred to mistakenly as an eight bit encoding, since microprocessor architectures at the time used an octet or byte, as many microprocessors continue to do today. In fact, seven bits are used for encoding, as already described. In a classic lack of foresight, the designers of the ASCII standard considered and rejected eight bit encoding as providing far more characters than the standard required or would ever require. In fact, eight bits means that in practice ASCII codes include an extra bit. The redundant bit travels with the code. It often functions as a parity bit for simple error checking. The odd or even value of the parity bit signals that transmission was successful or that the code is corrupted. The parity bit is a pure sign of digital existence, an ontological presence of ASCII in that other space of the net.

ASCII is less an encoding than a function. It does not describe the appearance of glyphs on screen. It does not prescribe type style, font, document structure, or markup. It does not specify how data will be recorded; it only prescribes encoding for interchange. The Foreword to the original 1963 standard states that ASCII is a "character set to be used for information interchange among information processing systems, communication systems, and associated equipment" and that the "means of implementing this standard in the principle media, such as perforated tape, punched cards, and magnetic tape" will be specified in later standards.[5] Additionally, Section 5 of the standard, *Qualifications*, states that this "standard does not define the means by which the code is to be recorded in any physical medium."[6] ASCII is a theory of the a-mediality of information.

What interests me in all this is ASCII as waste, or as "idiotic," in Clement Rosset's useful characterization of the real.[7] ASCII is immaterial but "pasty," as Gaston Bachelard put it of water.[8] This inert otherness poses the communicative or cultural questions around ASCII. This is the "ASCII unconscious," as Alan Sondheim puts it, a field of what is mine, what is appropriate-able and experience-able through ASCII. As with the Lacanian unconscious, ASCII is "like a language," a framework for symbolization, exchange, and proliferating partial encodings. In truth, nothing is encoded; there is only worked-over matter in and through imagined relations to otherness. As Lacan put it of the unconscious, ASCII does not "pose a question before the subject," since this would mean that we could come to the place of ASCII encoded information, precisely the place of our unconscious and being. Rather, ASCII "raises the question in that place *with* the subject, just as one raises a problem with a pen."[9] What is written in this place is the literariness of ASCII—which does not necessarily mean literature but rather becoming-literary.

Character and glyph

On the Unicode Consortium's "Acclaim for Unicode" page, James J. O'Donnell, the classicist, digital humanist, and provost of Georgetown University, declares, "Unicode marks the most significant advance in writing systems since the Phoenicians."[10] The significance and advance need to be followed and understood. There is only one code on the net and it is Unicode. Read the following carefully: it is "the universal character encoding standard for written characters and text." The Unicode Consortium's "What is Unicode"? web page, providing their central definition, begins with the following mantra: "Unicode provides a unique number for every character, no matter what the platform, no matter what the program, no matter what the language."[11]

From most points of view, the capacity of Unicode is enormous. "The majority of the common characters used in the major languages of the world are encoded in the first 65,536 code points," states the Consortium, but the sixteen bit encoding has the capability to encode up to 1,114,112 code points.[12] Long gone are the quaint old days of ASCII's 128 code points. Unicode even encodes fictional writing systems such as Elvish or Klingon. Its codespace includes any writing system whatsoever, without regard to whether this writing was ever employed by human culture. The standard declares, "more than sufficient for all known character encoding requirements, including full coverage of all minority and historic scripts of the world."[13] What an achievement! To encode all humanity's writing systems, past, present, and fictional! More than this, since streams of ASCII are the basis of all file transfers on the net, since ASCII is now a subset of Unicode, and since Unicode provides a structure for exchange and storage of data, then we must recognize that this encoding is the fundamental writing of all that is on the net.

Brian Lennon's excellent discussion of Unicode focuses on the limits of its claims to totality: "As a system defining universally efficient transmission, it must emit redundant waste in the form of merely local variation, precisely to claim its place as system."[14] As with the ASCII standard, Unicode is easily critiqued for the way it inevitably re-maps geopolitical concerns and contentions. But it is the distinction between character and glyph, central to the logic of Unicode, which accounts for the literariness involved. The enormous range of encodings, on the one hand, and the flexibility in transmission, storage, and display of characters, on the other, all come down to this distinction.

Joe Becker's 1988 draft proposal for Unicode, the standard that incorporates and replaces ASCII, states that a "clear and all-important distinction is made between characters, which are abstract text content-bearing entities, and glyphs, which are visible graphic forms."[15] This distinction continues in the standard implemented and maintained to this day. It is in operation within all encodings and implies a philosophy of the screen, of the object, of writing as waste, and of the literariness of digital inscription.

The Unicode 6.2 Core Specification released in 2011 states, "Characters are the abstract representations of the smallest components of written language that have semantic value."[16] What is meant by "abstract"? Character refers to the "abstract meaning and/or shape, rather than a specific shape." To say the least, this distinction between abstract shape and specific shape is rather abstract. What is meant by "semantic"? Not that a given character is meaningful. The character "a" may or may not be meaningful. Rather, according to Korpela's *Unicode Explained*, it "would be better to say that a character has a recognized identity and it may be sometimes used as meaningful in itself."[17] So, recognition as identity precedes meaning. Meaning and semantics are part of the recognizable identity of the character within the system of Unicode. The Unicode Standard "Technical Introduction" states:

> The character identified by a Unicode code point is an abstract entity, such as "Latin capital letter a" or "Bengali digit five." The mark made on screen or paper, called a glyph, is a visual representation of the character.[18]

Ferdinand Saussure writes that the "actual mode of inscription is irrelevant, because it does not affect the system. [...] Whether I write in black or white, in incised characters or relief, with a pen or a chisel—none of that is of any importance for the meaning."[19] Freud writes that "the single letters of the alphabet [...] do not occur in pure nature."[20] Character encodings, encodings of character. From here, we move to extreme rendition.

Extreme rendition

It is common to refer to character rendering. Characters are rendered on screen. To "render" is to express, to represent, to give, to produce, to surrender, to narrate, to vomit, to melt down and clarify the fat from an animal, to extract by means of heating.[21] It is a phenomenology of characters:

1   as inscribed intentional glyphs, however technically occluded the agency and pragmatics of intentionality;
2   as surfaces intensified through graphematic marking; and
3   as necessarily given to the eyes for reading subjects, however distant and mechanized those readers may be.

Take *rendering* as organic continuity, as an epigenetic landscape of the organism within technical display, as a flesh infrastructure. To render is to narrate: every rendering is a tale, a story of the flesh. To render up is to give and display before power. Rendition is the transfer or handing over of persons or property, as giving an account before the law, and *extraordinary* or *extreme rendition* is transport across jurisdictions, handing over, display, and disposal of bodies beyond the law.

My point here is character rendering as transcendence, or more properly the "overextended transcendence" Walter Benjamin taught us to melancholically read in the ruins of the baroque and their figuration in texts. Such over-extension should also be thought in terms of the numb narcissistic self-extension of McLuhan's understanding of media. Over-extension as necessarily a fiction, a positing, a sign, but also a pact before the law—in the sense of rendition—and such a pact is between communities of writers bound to the secret of plaintext.

Turn to Sigfried Gideon's profound *Mechanization Takes Command*. At the center of this book is a chapter entitled "Mechanization and Death," a meditation on the slaughterhouse, and on the history of machinic production as "the mass production of meat."[22] His descriptions of skinning, splitting, deboning, bleeding, and all the rest follow machinery's "adaptation to the animal shape," follow the production of meat at interface, follow the interface as a machine for producing meat, for organ-izing. Gideon writes:

> What is truly startling in this mass transition from life to death is the complete neutrality of the act. One does not experience, one does not feel; one merely observes. It may be that nerves that we do not control rebel somewhere in the subconscious. Days later,

the inhaled odor of blood suddenly rises from the walls of one's stomach, although no traces of it can have clung to the person.[23]

Note this final "no traces of it can have clung to the person." He insists "mechanization comes to a halt before living substance." He re-directs us from the traces of touch, as in Benjamin's investigation of the haptic dimension of cinema, to an impossible interruptive transcendence. This is perhaps my only point: our body interrupted at the machine as the transcendence of our body. The literary scene as our body before the neutral screen, turning by the odor of other bodies, yet bereft of any material trace.

Once again, Becker's Unicode standard states that a "clear and all-important distinction is made between characters, which are abstract text content-bearing entities, and glyphs, which are visible graphic forms." What is a glyph? A glyph is perceived. It is phenomenally given in the world of appearances. It "represents the shapes that characters can have when they are rendered or displayed." It is technically produced. "In contrast to characters, glyphs appear on the screen or paper as particular representations of one or more characters."[24] Glyph and character are part of a single writing technology, the one side abstract, encoded, and conceptual; the other side, material, perceived, and undefined.

Glyphs outnumber characters. There are multiple but finite possible renderings for a character: many sizes, many resolutions, many forms of visibility, any of which may be recognizable as characters. Glyphs are recognized for the encoding rendered but are fundamentally unnumbered. They appear and are seen but not defined by Unicode. Appearances are disordered and without accounting. The difference is between wild phenomena in an entropic waste field, on the one hand, and specific forms or characters, on the other.

The character on screen, the text that I see and read, is not a body. There is no character, only encoding read through the appearance. What I see, but not what I read, is an innumerable disordering of appearances. Characters are the logistical, weaponized construction of "reading" and "seeing" through systematic technical distribution of the visual and symbolic. There is capitalization of characters through blind sloughing away of the matter of glyphs.

Marvel with me at the wonder of glyphs that appear and vanish! The flowing and fleeing energetics of the screen are mobilized and lost by biotechnics that insist on the character–glyph distinctions, and that hold fast to character encoding into ASCII and now Unicode.

The history of character and glyph are similar to each other. The earliest English usages of both refer to a carving, a cutting, or a marking. In short, usage refers to the problem of rendering. By the seventeenth century, character came to refer to the mental and moral qualities of an individual, while glyph continued to refer to the material and physical engraving. As McLuhan detailed in the *Gutenberg Galaxy*, the emergence and dominance of print led to the perceived metaphorical equivalence of the qualities of individuals and printed characters.[25]

I tell you: not just a letter but also any appearance on the screen is a glyph etched or inscribed on that surface. Every screen is a glyph; every screen is the *stochoie* or *littera*. The literal is always being screened. Screen refresh is about twenty-four flickers per second, or one thousand forty images a minutes, 86,400 images an hour, more than a million a day, a quarter of a billion a year, bursting on screen and flowing away, gone. And that is just one screen. Our machines are dedicated to this tremendous and ecstatic expenditure of images.

Character encoding means *I write it, it writes me*. I wrote this. To screen is to touch, is to contact the other. But the other never touches such a terminal screen; the other is never smearing the same fluid on my fingers, on my tongue. How rigid is this "never touches"? What relations exist through this rigidity?

The answer: a literal relation through the explosion of the rendered glyph. Unlike the abstraction of coding, which happens in some other light, in the great beyond, the letters with me here become nothing, there is no becoming, the letter is and is destroyed, it is over-exposed, fading in screen flicker. Every character is a discharge of energy, a wasting of my body, a part of the continual necrosis of the flesh in this world of light, in a world of entropically decaying suns.

Plaintext performance

A specific work entitled "Plaintext Performance" by Bjorn Magnihildoen appears in the Electronic Literature Organization's

Electronic Literature Collection Volume 2. Here's the useful author's note:

> It's a live writing performance over the net combining (1) keyboard writing, (2) machinated, algorithmic writing, and (3) feeds from the processes surrounding the writing (like system monitoring, net connection monitoring, ftp log, etc.). All in realtime and plaintext. It was performed live at the BIOS symposium, Center for Literary Computing, West Virginia University, September 2006, with a unix ytalk session as a sideshow. The static version shown here is based on the exhibition *e and eye—art and poetry between the electronic and the visual* at Tate Modern, London, October 2006. It's part of a series of work called "protocol performances." "Protocol" is meant both as a lower level set of rules of the format of communication, and as statements reporting observations and experiences in the most fundamental terms without interpretation, relating it to phenomenological "noemata"—thought objects, and thus identifying a data stream with a stream of consciousness. "Performance" is meant both as a data protocol's physical performance as much as its play on the meaning as an artist-centric execution of work.[26]

First remark. Such descriptions of pieces of electronic literature are essential. The literary critical act of description is at least as important as the work. For every performance, there is description. The artist is always asked, how was it made? The work comes with this demand. Every performance or reading I know of, every published work, is wrapped and explained technically, especially when it comes to electronic literature. It is excused, an excuse given for how it was made. The literary is somehow in this paratextual explanation, the interest of the work tied to the author or critic or author as critic's discourse about it.

Second remark. Plaintext performance is problematic to describe. Magnhildoen calls it "live writing performance over the net combining" etc. but what I see on the screen is a very long file. The performance is elsewhere from the explanation. Of course, I know it was performed and in a particular place and time. I should know, since I commissioned and curated the performance as part of a conference I ran entitled "BIOS," as the author notes. This knowledge is particular to me and my friendship with Magnhildoen. When was the performance? This seems easy enough: it occurred on a screen in a room at West Virginia University. It also occurred on a screen somewhere in Norway, where Magnhildoen is located. It also occurred—if that is the right term—on a series of data packets exchanged between servers. I know, and am reminded by the "previous publication" information, that the version we are looking at is not the one from the WVU BIOS conference but from a month later, "Monday 16 October 2006" at the Tate Modern Museum in London, in the "e and eye" exhibit, where I again invited and curated Magnhildoen, in collaboration with Alan Sondheim, and where Sondheim and I had been invited by John Cayley and Penny Florence, who organized the whole thing. All of which is to say that the performance was in multiple prior times and localities, and with many acquaintances and friends; indeed, I'm not sure now I can call Magnhildoen a friend, this man I have never met.

"Plaintext performance" is live writing but at several times and in or through several lives, living on in many ways. Plaintext performance, this one, this work, but also the general concept or what I think is a concept, relates to this multiplicity of friendship and writing.

Third remark. The ELO collection editors' description adds to the complexity of reading this work:

> At once nearly cloistral in its impersonal sparseness, a black fix-width typeface against a white background, revealing nothing of the author's thoughts or dreams, animated by a nearly-trivial algorithm and on the other hand baroque in the complexity of what is in the text file, ASCII art, various computer outputs, garbled poems, mangled transmissions, and random text/number strings that suggest a deviant psychology, this minimal "performance" contains just enough life to bring us back to the earliest fascinations with computer sentience. For what little the program throws up against our "reading," it is just enough to remind us of the Other that is the computer algorithm.[27]

Now, it is this series of qualifications and partial concepts, on the part of what I earlier described as the necessary literary critical wrapping that accompanies and situates electronic literature, it is this undecidability that qualifies plaintext as the minimal criteria for the literary: it is "nearly cloistral," "revealing nothing," a "nearly-trivial algorithm," "on the other hand baroque," "just enough life," "just enough to remind us of the Other," and of course a "minimal 'performance,'" where the word performance is in quotes, that is, textually performing performance.

Fourth remark. The ELO editors point to the text's animation by a "near-trivial algorithm." The text inches along in a charming old school staccato manner. What I cannot figure out is the following: Is this animation the performance? If so, the text is not read and its movement is the point, displaying obtusely and conceptually the relation of text and code. Or is the text the performance and the animation a way of showing the text, the animation added to the plaintext performance, plaintext performance being performed, a doubling of the performance, doublings which we deal with constantly. We could see the most complex multimedia, mixed reality, what have you, as the edge of plaintext in performance. Performance means this separation, the synecdoche of plaintext and performance. Performance as an excuse, as if to say "sorry for the work." It moves, it shakes, it must perform. It means the performance of nothing but performance.

Performativity is one of a familiar chain of concepts or rhetorics explaining the auto-affective looping of subjectivity and knowledge before networked screens. Magnihildoen's "Plaintext performance" is one of those famous "technotexts" described by Kate Hayles.[28] I use Hayles' terminology only momentarily: it too is part of the chain of concepts, along with rhetorics of mnemotechnology, semiotics, and so on. I am purposely being dismissive and problematically emphasizing rhetorics to insist on such concepts as a ways of deepening text through allegories of apparatuses and beyondness caught in the apparatus, allegories providing knowledge of this capture though not of the beyond, ways of construing an inscription that both negates itself, erasing the matter and economy that marks it, and at the same time performs this erasure, inscribing inscription inscribing, and so on. The result is melancholic knowledge, melancholy or traumatized mourning as yet another concept in the chain. It is we who share in this knowledge, we literary critique, one and all, we are organized, given organs, formed into a crowd in the process.

Apologies, but there is another, fifth remark. Part way through the text, a quoted remark, without any attribution reads:

> Well for a start, the dreaming, maybe obviously, of this plaintext performance as ecstatic text, and as a performance with shamanic points, the spirits would be the machines so to say, routines and memory here, the underlying system of PP which uses the server for read-write-rewrite operations quite autonomously.[29]

Is this meant to be about the work? Ecstasy and shamanic inhabitation as transcendence of my remarks and theories and questioning. I think of the Shaman religion of the Sami people of Magnihildoen's Norway, mediators between earthly and spiritual. Plaintext performance as dreamtime transcendence. Sorry, but to show this, I must return to cryptography.

One time pad

William Friedman—director of SIS cryptanalysis, technical advisor at the National Security Agency, from its founding by an 1952 executive order that was itself immediately classified to hide the fact of the agency's existence, inventor of the M-134 precursor to the military's ECM Mark II or SIGABA, a multi-rotor proto-computer encryption device with pseudo-random operation—wrote the book on American military cryptography. But all this was a warm-up. In 1955 he retired, with hefty hush money payoff from the government, and focused his cryptanalytic skills on the authorship of the Shakespearian plays. He and his wife Elizabeth were "Baconians" since before the First World War, and with techniques developed by and for the NSA—then and now the largest employer of PhD's and computer programmers in the world—they set to work on Shakespeare. Their book *The Shakespeare Ciphers Examined* appeared in 1957 and won the Folger Shakespeare Society Award.[30] Despite this honor, the book failed as a decryption of Shakespeare. Unfortunately for the Friedmans, the cryptanalytic science called "the most potent secret weapon of World War II" fell short when it came to literature. No cipher could be definitively shown in Shakespeare's work, though no end of decipherment was possible. The length and complexity of the material exceeded the limits of rigorous, final decryption, leaving only partial decodings, that is, leaving literary criticism. The sheer literalness of plaintext gave away no secrets.

Meanwhile, from April to October 1960, Jackson Mac Low composed *Stanzas for Iris Lezak*, in part while riding the subway between the Bronx and Manhattan.[31] The poet's early works were expressive and surreal, but by the 1950s he was composing using chance operations—flipping coins, throwing the I Ching, and so on. In *Stanzas for Iris Lezak* he first adopted the "chance acrostic" method that characterizes much of his work up until his death in 2004. Mac Low selected an "index sentence" and then spelled out each poem acrostically from the index sentence, using words found through pre-determined operations. These operations varied from poem to poem, but typically involved selecting a book or magazine, and then reading until encountering a word beginning with the first letter of the index sentence. This provided the first word of the poem. Mac Low then read on until he encountered a word beginning with the second letter of the index sentence, providing the second word of the poem. The process continued iteratively using the index sentence to generate the poem from the source text. Index sentences were chosen on impulse and embedded personal choice in procedural text generation. The first six poems in *Stanzas for Iris Lezak*—"6 Gitanjali for Iris"—use what Mac Low called an "exultant" index sentence, a macho dedication to the poet's girlfriend: "My girl's the greatest fuck in town. I love to fuck my girl." Subsequent poems generate the index sentence from the source text itself, using the title of the text, the first sentence, or some other text string. Poetry no longer floats like Wordsworth's cloud or offers many ways of looking at Stevens' blackbird. Instead, poems are found like the following, from *The American Heritage Word Frequency Book*, a source for many of Mac Low's works, at frequencies between 1.0623 and 1.0611 for U (word frequency per million): "fortnight hues weaknesses scaled cilia heaven's soberly closeness nobody's laughing lessened Vincent."[32]

Of course, Mac Low used exactly the method Claude Shannon used to represent the probability structure of an information source in 1948's *Mathematical Theory of Communication*. "Pattern Recognition by Machine"—one of Mac Low's first chance acrostic pieces, which begins "Perceive. As Letters. Think? Think? Elusive, relations, now met most of the classic criteria of intelligence that skeptics have proposed."—works through a 1960 *Scientific American* article on information theory.[33] Aleatory content cites a cultural milieu of circulating discourse on the informatic structure of language. Admittedly, the pattern and resulting meaning of the poem are procedural outputs. The procedure empties the citation and proves the point of the citation in doing so.

And, of course, Shannon's work was precisely enabled by Friedman's cryptography. Friedman's 1920 "Index of Coincidence and its Applications in Cryptography" was the first attempt ever to formalize the statistical regularities of letter and digraph or letter pairings in English.[34] An itinerant knowledge of frequencies existed in cryptographic texts from the first, with Lavinde's famous *Trattati di Cifra* of 1480 containing one of the earliest letter frequency tables. This knowledge was materialized in the famous sequences of lead slugs "etaoin shrdlu cmfgyp wbvkxj qz." The sheer repetition of these letters left them in linotype boxes until Friedman connected frequency to the structure of the language. In doing so, he dissolved the lead slugs into one aspect of a schema of mathematical frequencies. Friedman showed that any two consecutive characters selected from a plaintext would be the same character 7 percent of the time on the average, due to the common nature of digraphs such as "ee" or "oo" or "tt," whereas in purely random text the same character would only occur twice 4 percent of the time on the average. Unlike a randomly generated string of characters, a message will possess frequencies that persist as traces in even the most complex encryption. The 3 percent difference between 4 percent and 7 percent made all the difference: just this was sufficient to render cryptographically insecure any key generated from a handy book, the traditional source for random keys. Shannon's information theory provides a mathematical explanation of this "index of coincidence." As a result, every information source is modeled on cryptography.

Shannon, by necessity, exemplified information using language. His famous elaborations of how a probabilistic selection of English language characters "approaches a language" is not simply an example of informational probability but the exemplary instance that models all others. The probabilistic structure of any information source will resemble this model. Shannon constructs "typical sequences in the approximations to English," ranging from "zero-order," a string of "symbols independent and equiprobable"—

which reads something like this: XFOML RXKHRJFFJUJ ALPW XFWJXYJ FFJEYVJCQSGHYD QPAAMKBZAACIBZLKJQD— to "second-order word approximation," where "word transition probabilities are correct," that is, where the likelihood of one word following the other reflects the structure of the language but where "no further structure is included."[35] The example reads:

THE HEAD AND IN FRONTAL ATTACK ON AN ENGLISH WRITER THAT THE CHARACTER OF THIS POINT IS THEREFORE ANOTHER METHOD FOR THE LETTERS THAT THE TIME OF WHO EVER TOLD THE PROBLEM FOR AN UNEXPECTED.[36]

Shannon insists that ATTACK ON AN ENGLISH WRITER THAT THE CHARACTER OF THIS "is not at all unreasonable."[37] What is reasonable about it? It is as if the "frontal attack" were a violent thematization of the disjunction of information and meaning. Shannon does not comment. Of course, the string is nonsense but it is given as an example, the only example, of "second order word approximation." The string is an example of the literalness of information and voids any apparent meaning in making this example. Information means the production of this string. Shannon adds that in the sequence of examples, the "resemblance to ordinary English text increases quite noticeably."[38] What is resemblance here? The heuristics of resemblance are the background aesthetic, the necessary opening on which electronic literary meaning is built. The presupposition of an aesthetics of resemblance is built on the positional force of rhetorical circuitry, the material turn of words where program and reading meet. "Not at all unreasonable" means sufficient for worldwide networks of open secrecy.

Shannon's method was to compose streams of characters with a book of random numbers, in conjunction with a table of letter frequencies. Word order approximations, that is, where the information source represented contains complete words in the English language, become exponentially more difficult and Shannon turned to an easier method: he selected a book off the shelf, opened it at random and selected a word. He shut the book, re-opened it and selected another word for the next in the example, and so on. For second-order approximation he selected a word and the word next to it. Word to word transition in printed text was an adequate representation of the statistical structure of the entire language. Moreover, the corpus of printed texts represents the superset for statistical analysis of the language, as word frequency books show—the *American Heritage* 1971 version is a "computer assembled selection of 5,088,721 words (tokens) drawn in 500-word samples from 1045 published materials (texts) selected from 6162 different titles."[39]

Mac Low adopts the same techniques and names the results poetry, foregrounding the surplus at work in the theory of probabilistic information sources. What Shannon's theory, as exemplification or model, names as given in every information source—truly given, truly a gift, a poetic gift—is precisely this surplus. The vogue for "codework" literature—such as Magnihildoen's—reflects an awareness of this gift. Even without following the links from Mac Low's Fluxus-associations to the contemporary codework and net. art scene, it is clear that such works present the crux of information, in a repetitive but always striking way, reminding us that what we see is not what we get, and forcing the experience of negotiations between seeing and getting. As I argue elsewhere, it is not necessary for the code to "work," at least in the strict sense described by the institutions of computer science and machine construction, that is, not necessary for the code to be compilable and executable on a standard microprocessor running an OS that produces a finite output.[40]

"Logical depth," as developed by Charles Bennett at IBM, is increasingly adopted as a standard in network administration, digital security, and other fields, to measure the complexity of outputs—of whatever sort: numbers, images, etc.—qua underlying algorithmic processes. Bennett proposes depth as a "formal measure of value." In the background is a notion adapted from Turing's theory of computation, which takes the complexity of algorithmic output and the noncompressibility of algorithms as models of complexity in general. Complexity is value. Bennett writes:

> … value of a message thus appears to reside not in its information (its absolutely unpredictable parts), nor in its obvious redundancy (verbatim repetitions, unequal digit frequencies), but rather what might be called its buried redundancy—parts predictable only with difficulty, things the receiver could in principle have figured out without being told, but only at considerable cost in money, time, or computation.[41]

By contrast, reversible computing or thermodynamic computing asks whether the energy dissipation in running a computer ultimately balances out with the computation produced. The simplest operation of AND or OR, even at the assembler level, involves destruction of information and expenditure of energy. Copying data would seem to increase information and deleting data to reduce it, but this is not the case. The computation involved in deletion still necessarily expends energy. So, tallying up the balance becomes complex. Following Landauer's Principle, if computation were shown not to expend energy, the result would be a time-reversible process, a violation of the second law of thermodynamics, and the triumph of the digital over the analog. Computation would be a net gain, the super-production of immaterial, timeless, and virtual information. It turns out not to be the case, despite the near-reversibility of today's processors and despite evidence for reversibility at the quantum level. Before there is digital simulation, there must be flow. The mere need to power the computer, to flow electricity into the material shell that simulates the digital space, establishes the energy spent and the non-reversibility of computation.

My laptop is scarred and stained and chipped. Its solidity is a trap for time. I touch the screen. The cool surface is a non-revsersible gradient of energy dissipation. Boron, arsenic, phosphorus, antimony: these dopants are impurities added to a semiconductor lattice to alter its electrical properties. The resulting shift in the Fermi level of the semiconductor makes the semiconductor into a storage medium. Trace radionuclides generated from the ceramic packaging used as the substrate in computer chips can change the contents of a computer memory unpredictably.

All information tends toward the material singularity of the cryptographic ideal of a "one-time pad," a simple and supposedly unbreakable cipher. In general, cryptography encrypts plaintext messages by combination with encrypting texts following some algorithm. The most immediate way to decode a cipher given only the encrypted text is to search out regularities or redundancies, clues to the key and method used. A one-time pad demands a random key of exactly the length of the plaintext. The random key possesses no internal structure, no redundancy, while the fixed length prevents any repetition, a classic error offering mathematical clues about the key. The unbreakability of the one-time pad is its algorithmic complexity: with no redundancy at all, the time spent reconstructing the plaintext from encrypted message by pure guesswork is prohibitive to the point of impossibility. One-time pads were in use by cryptographers since at least Vernam's patent of the method in 1917. Its mathematical formalization as a "one-way function" formed the basis for the Stanford Digital Encryption Standards (DES), now in place for all digital communication.

Of course, the unbreakability of the one-time pad requires its proper use. Most of all, this means an absolutely random key. As Friedman points out, books were the traditional source of random cryptographic keys. Just pick up your favorite novel for a source with sufficient randomness. The one-time pad seems to end this method, though digital encryption consultants still advise us to type a "favorite poem" when generating random bits for PGP encryption. Better yet, "spice" up the poetic phrase with informational redundancy—that is, create a pseudo-random string following Shannon's practice of itinerant codework. Resist Security offers the not unreasonable password "Willy Wonka meets the Terminator" and spices it up to "willywoNKA meet$ the^ terMIN@tor." In strict accordance with Friedman, they add, "NEVER WRITE YOUR PASS PHRASE DOWN!"[42]

Nevertheless, the one-time pad would seem to require a different complexity than that offered by texts. The dream of contemporary cryptography is using quantum sources to produce random keys. The limits of digitally generated random sources are the on-off states of bits. The complexity of such a source is limited by the combinatory possibilities of the bit-set. The wave-particle duality of quantum sources, by contrast, means that quantum bits can be in multiple states, 1 and 0 at the same time, oscillating at 32-bits or 4,294,967,296 values in a single calculation. Rather than generating a random key from the pulse of the computer's clock, the typical source for pseudo-random number streams, quantum cryptography generates keys from light, using the polarization basis of photons in a fiber-optic cable. A leading technology in the field is MagiQ's Navajo, a product name that cites the Second World War use of the improbability of Native American languages as a cryptographic technology.

The premises of "traditional cryptography" assumed a single key, however random, and thus a single plaintext. Quantum-generated keys destroy this self-evidence by creating an oscillating number stream with no regular states. With the key and the encryption process no longer accessible to the senses, quantum cryptography would seem to bring about the end of the "textuality" of information. In fact, the uncertainty and inaccessibility of the quantum is not simply the end but apotheosis of the book, the seemingly final exemplification of the complexity of the literary. Quantum cryptography is an infinitely deep source of decryption, just as Shakespeare was for the Friedmans. The point is not that quantum cryptography is in some way literature but that the "form" of its complexity is modeled on the complexity of texts and best exemplified by literary texts.

Friend request

Leaving us with plaintext. Let me re-emphasize that the term "plaintext" originates in cryptography, prior to the computer and prior to lowest-common denominator ASCII text and the like. If plaintext is a message to be encrypted, and ciphertext is the encrypted message, with some algorithmically determined mixture of transposition and substitution ciphers, there is also cleartext or a message transmitted without encryption. For cryptographers, cleartext is a message "in the clear," readable and understandable by any receiver.

How does plaintext differ from cleartext? Indeed, the two are frequently used as synonyms, even in a canonical pre-computer work such as Friedman's *Elements of Cryptanalysis*, written in the 1920s and long the standard teaching text for U.S. Army and later NSA cryptographers. Sender and receiver must possess the same one-time pads to encrypt plaintext. Typically they share instructions, something like this: "encrypt the next message with the next one-time pad." Once a message was encoded, writers ingested and swallowed the one-time pad to avoid its capture, or one-time pads were printed on nitrocellulose to self-destruct after usage, *Mission Impossible* style. Again, plaintext is a domain of matter and bodies smeared across information systems. The randomness of the one-time pad combines with plaintext to form the messages, indeed all the messages in all information systems.

The one-time pad was given its perfect formulation in 1949, in Shannon's "Communication Theory of Secrecy Systems," an essay as important as the previous year's *Mathematical Theory of Communication.* Shannon shows that the one-time pad mathematically ensures perfect secrecy, which is to say, the secrecy of every message in every system in all the networks we deal with, which is to say, secrecy as the bond of a community of writers, a bond of intimacy that is inseparable from the literary.

Shannon's starting assumption is that "the enemy knows the system being used." He continues, "he knows the family of transformations, and the probabilities of choosing various keys."[43] This became known as "Shannon's maxim." It reformulates an 1883 argument from Kerkhoff that a secure crypography system must not be secret; it "must not require secrecy and can be stolen by the enemy without causing trouble."[44]

Why is this important? The maxim defines contemporary information systems, systems with no interior where everything is in the clear and open, systems of absolute disclosure and display. Information is secure by being complex to access, rather than physically hidden. There is no security through obscurity. Everything is displayed but not everything is easy to display. There is an interiority secreted in information. If *The Mathematical Theory of Communication* describes the circuit between sender and receiver as premised on passage of message through channel, "Communication Theory of Secrecy Systems" theorizes secrecy of the message in light of the enemy's gaze, as it were. Think here of the absolute other of Levinas, already facing us, taking us hostage. Or think of the Googlization of language. This is the condition of all information: the enemy has already stolen the system.

Where "the enemy know the system," perfect security is defined by the literary secret of plaintext. Reverse "the enemy knows the system" to the system is knowable because of the enemy. The system is made discrete, made functional by the other, who witnesses the exchange, sees the message in the clear but cannot decrypt it. Information finds an enemy, just as it posits sender and receiver, that loving pair of friends. You write to me. I receive your message. Only I possess the one-time pad to read it. What is this pad? An intimacy already in my interior, an inhabitation that we share. Friendship is this possibility of reading the other's messages. The circuit of information—I send to you, you receive my message, and so on—is a fraternity, a bond, and a community. This is the interior of information that displays itself in everything that passes as communication. This pertains to read-write-execute permissions and handshaking protocols and logins and CAPTCHAs and spam, as I have written above, and pertains also to practices such as file sharing and piracy and open sourcery. Think of the digital semiotic *chora* that Julia Kristeva identified with poetic language.[45] Think too of the noematic protocol statements that Magnihildoen identified in the strata of plaintext performance.

A secret remains at work in plaintext. Not obscured, not there to be found. It is in the open, in the clear. The secret of friendship is not in the plaintext that stays locked in a safe, nor is it in the one-time pad, burnt or eaten. Plaintext posits a topology of materiality and information where I am already intimate with the sender, already a friend, we already share one-time pads, already share objects and organs that are external to the system. I tell you this is the most important thing: from the singular consumed material of the plaintext message, to its encryption through the contingent one-time pad, into plaintext as ciphertext, into every text we read today. In all this, a secret is contained within the texture, the literary text as this secret. The secret is the interior of the subject in information, a secret shared by communities of writers but created poetically by the other, by the enemy, who knows the secret exists and who forces us to be secure, that is, to be who we are as secure.

In actual systems, in actual settings, are we friends? Is it not the case that there are always messages between friends, as well as messages that I receive and hope are from friends? Is this not the possibility of literature? Is this not the love of literature that I invoked already? Which is to say: to read messages as literature is to take the risk the messages are from a friend. All messages are this way. You and I here, when we communicate, when you read my book, or when I email you, or when I chatroullette and pop up on your screen, or when I friend you on Facebook, or all the rest. Are we friends? What friends are there in writing? Which is to say, what enemies? Apologies again, there are no answers to these questions. Let me say I can only apologize to a friend. I can only say sorry to a friend. Apologies are only possible when there is a message, a secret we share. Sorry, my message ends here; only you, my friend, can read what follows.

IX

Bodies never touch

Look at *Second Life*, that free 3D virtual world where you can "experience endless surprises and unexpected delights in a world imagined and created by people like you."[1] Look at Alan Dojoji or Julu Twine, never touching, but sometimes engaged in a stylized and disordered dance or combat with other avatar toons such as Sandy Taifun. If you were logged into Second Life and looking at the scene through your own avatar eyes you would read the avatar movements as odd, nothing like human motion. Alan Dojoji is probably in a particulate haze emitted from the avatar surface; somewhere in there is the toon body, ascetic and sadhu-like. The avatarts move at their own speed following a Virilian logistics of perception, but their movement is intensely personal and utterly subjective. Second Life artist Alan Sondheim (a.k.a. Alan Dojoji) calls it "inconceivable." You can mobilize a story around the avatars. You can cluster the signs and read a narrative. The avatars quarrel and fight; they dance a *pas de deux*; they engage in a ritual mating dance; they socialize and connect.

Their avatar dance is a stotting, and a provocation to reading and narrative. If we agree that *Second Life* is their scene, if we agree that this is a performance within *Second Life*, if we agree to call it a dance, then certainly their drama is being witnessed. There is always an audience: even while artists Alan Sondheim and Sandy Baldwin are directing their avatars Alan Dojoji and Sandy Taifun, they are watching their avatars move, they are watching the performance taking place. More often, there are others watching as well. The audience may wander in and out or stare transfixed, but they are there nonetheless. You can watch the dance on YouTube and Vimeo.[2] The apparatus contributes as well, of course. The chat is logged, the video recorded, and the sysadmins at Linden Labs are in the background, too.

The welcome to *Second Life* reads: "We hope you'll have a richly rewarding experience, filled with creativity, self expression and fun."[3] Mark all these terms: "we," "you," "richly rewarding experience," "creativity," "self expression," "fun." Codes with addresses, signals with sources. *Second Life* coheres as "world" through a semiotics of theater, virtual world as stage for delegates of human actors and the desire for fidelity to the real. Enormous work is expended building replicas of real structures and bodies.

The story goes that the movements of avatars in *Second Life* are directed by codes, by precise protocols. The movements are puppet manipulations performed by a user or a script. The subject of the avatar is absent. Better, set this in an ontological situation: the avatar is the transcendence of the subject and the body. It is an epistemological image of surpassing embodied subjectivity. By this, I mean the following: to operate in this field, to use and view an avatar, is to know the representational outcome of the embodied subject as nothing more than a preferred image, something wagered or hypothesized. Such an image is purified, cleansed, and detoxed of subjectivity and embodiment. The experience of an ego beyond the protocols and codes collapses and shrinks to no more than a narrow channel of giving the image, a giving that is shown by the selection and display of avatar bling, from clothes, to body parts, to cool dances and gestures.

What control of the self! What mastery of the body! What a stiffened, contained, and finely held existence! Or so the story goes.

The avatar is the visible sign of an announcement and assertion of presence of a user that operates the avatar as a puppet. The sign is read by other avatars and by their users. This semiosis of the avatar body is a familiar one. Take Lisa Nakamura's well-known argument in *Cybertypes* that race is "written into" online representation even when it is not made explicit or "declared," as one might say of a data type. This leads her to declare the urgent need to

> direct critical attention to the conditions under which race is enunciated, contested, and ultimately suppressed online, and the ideological implications of these performative acts of writing and reading otherness.[4]

The conditions of critical attention require taking the avatar as a sign.

When an avatar gestures and dances in *Second Life*, when a chatterbot chatters and spews text, it is possible to see these artificially created events as vitiating the connection of words to bodies. If words can be produced, generated, endlessly chattered, then there can be no intimacy in and through words. If the avatar is self-mobilized and if its gestures are copied and coded, then there can be no contact between avatar and self. Following Nakamura, we can look for the suppressed ideological contents written into avatars. We can read and understand the avatar in terms of cultures that are elsewhere and that proffer the avatar, that propel it in and through the virtual world. The dance of the avatar is the dance of culture. To read the field of *Second Life* for signs of the absent subject is to see the virtual world as characterized by negation. Every appearance in *Second Life* would be the negation of the real, every avatar and object the not-real sign of otherness elsewhere. *Second Life* becomes the unconscious of first life; *Second Life* is "like a language" that articulates the truth of first life.

No doubt *Second Life* contains many languages, citations, vast circulations of text, but is it like a language? Should *Second Life* be understood through performativity and enunciation, through "writing and reading otherness," as Nakamura put it? Is otherness writable and readable? Let me repeat: if there was a scene, if there was a performance in *Second Life*, if we agreed to call it a dance, then certainly the drama would be witnessed. Does this mean reading takes place? Or is *Second Life* unreadable? The discourse around avatars emphasizes acts of representation and performance of the subject delegated to the avatar. But nothing is represented in *Second Life*.

Pervy intimate avatars

These are pervy avatars. The perversion is the return to the symbolic domain with the pleasure and power of the theatrical staging of the object, with the reflexivity of a theory of perverse desire in *Second Life*. However, this is not classic psychoanalysis where perverse behavior is a deviation from symbolic protocol. *Second Life* is a constant turning and inversion, a perviness in practice that thickens images, that wears down and smooths all things to paste or "prim"— that *Second Life* term for a simple geometric shape. Perversion in the sense argued by Janine Chasseguet-Smirgel, in the archaic wish for smooth forms, for undifferentiated sticky immersion.[5]

My avatar is the extension and completion of my body, the tip or terminal point of skin. I give it displaced and dismembered body parts: my eyes that see its world, my fingers that guide its movement, and my tongue that silently utters its words. Therefore, there is no "truth" of the avatar: not a more hidden self, somehow on this side of the screen or on the other side of the avatar, not a choice between an embodied or projected subjectivity. The avatar is part of the continuous excessive and extruded production of my body, filled with fluids and blood and new undiscovered organs.

I capture here, in this text, what occurs there, in the virtual world. The tableau of this writing, the narrative, is equivalent to the tableau and events taking place there. Something immobilized here, held stiff, sticking out for you to grasp: my text. A phallic remnant captured and contained from *Second Life*, a dictation from one inscriptive domain to the other. Ideally, I pose for you, you are satisfied with my writing, and I give you all I have. The material here is an organized theoretical machine that hums and runs, processing *Second Life* for you. You plug into the machine and get off on it.

The avatar is a single figure but many bodies inhabit it. To work with Alan Sondheim on this art means being jacked in with old-school motion capture equipment, where long spaghetti wires run from sensors on your body to large electromagnetic field generators. As you move, the mocap software registers displacements in the field as a moving array of points. The array is mapped onto nodes: left wrist, right hand, back, and so on. The sensors may be re-distributed: wrist is now stomach, hand is now forehead, your entire body transmuted.

There may be several of you sharing the sensors. The single, normative body map is stretched, folded, and messed up. Picture dancing bodies, wired for mocap. Standing there in person, you can read the tableau as ritual encounter and mating dance. In such settings, in the approach and engagements of eroticism, we imagine bringing our bodies and our selves as subjectivities and corporalities to share. However, the sensors do not record this part and that part of my body. There is no inscription of two subjects coming together for the symbolics at work in ritual. For the sensors, there are clusters of floating dismembered parts. Is this not the case in any encounter? Not for the viewer but for the dancers! The tableau dissolves not into a phantasia for the viewer to read off minds and bodies in harmony, but into a sticky fluid and secretive *no body* that is only for the dancers.

There may be no body for the sensors: you are not there, I am not there, and the sensors are swung, dragged, or thrown, dropped on the floor, dangled in the air. No body in the apparatus swarms of bodies. In truth, this is only one way that Sondheim works. His project crosses first and second and multiple lives. I can tell the history of the work in terms of visits and collaborations at West Virginia University's Virtual Environment Lab, with sponsorship and support from the Center for Literary Computing. However, I could also tell a history of longer duration, reflecting Sondheim's projects and concerns dating back decades.[6]

Rather than focus on this history, I can describe the apparatus. The captured motion is in a biovision hierarchy file, or .bvh format. The file collects arrays of node coordinates defined something like this:

```
JOINT LeftHandPinky1
{
OFFSET 9.34247 3.34117 -0.23928
CHANNELS 3 Zrotation Xrotation Yrotation
JOINT LeftHandPinky2
{
OFFSET 3.2826 2.50999e-005 0
CHANNELS 3 Zrotation Xrotation Yrotation
JOINT LeftHandPinky3
{
OFFSET 2.1292 0 0
CHANNELS 3 Zrotation Xrotation Yrotation
JOINT LeftHandPinky4
```

```
{
OFFSET 1.78389 0.0301608 -0.0105222
CHANNELS 3 Zrotation Xrotation Yrotation
End Site
{
OFFSET 0 0 0
```

It organizes the nodes accordingly in a temporal scansion, something like this:

```
MOTION
Frames: 1937
Frame Time: 0.0333333
30.1907 93.0362 244.596 -104.469 2.65601 110.943
-180 -4.2965e-031 -90 10.902 26.2596 15.6978
15.192 54.3653 31.084 15.1922 54.3657 31.0838
-27.2648 -3.24233 -5.51396 8.51242 -6.98865
65.9159 120.092 75.5343 58.4939 15.7529 -20.2383
-5.4426 -167.277 84.4999 48.8109 -78.5553
-11.0668 16.9617
```

This is a tiny example of an enormous file. The scansion can be stretched and edited, the motions mapped to other motions, the joints and nodes edited and renamed. The .bvh file can be rewritten, the textual editing driving the animation. In turn, the file is readable by many character animation programs, from 3D Studio Max to Maya to Blender. The .bvh files import into *Second Life* and drive avatar gestures. The dance is a dance with code and its structures. The heat and flex of bodies, distant echoes of motion, tether to execution and coding. The corporeal and technical are in disarray, neither separate nor together; many bodies, many motions, many scenes, many codes. All transcribe into a single button in *Second Life*: a single mouse press triggers gestures and gestural series. Culture in *Second Life* is managing bodily swarms; to read this culture is to make the swarm a sign.

A loose assemblage of apparatuses enunciates the subject in *Second Life*. This is the *Second Life* "viewer." The default is the Linden Labs Viewer but there are third party viewers as well. You run the viewer to connect to the world. Within it, you typically watch in a modified third person perspective, over the shoulder of your avatar. You can change to mouselook view, more distant third person views, and so on. You experience *Second Life* within this frame. What takes place follows a style dictated by these apparatuses. The resulting descriptions—my avatar did this, I do that, bolstered by screen shots and in-game movies—gather the diegetic products of the apparatus into a single narrative. The apparatus announces the subject as avatar. You enter the world, you encounter other avatars, you befriend others and interact with them, you chat, you gesture, you dance, and you build.

There is the body of the avatar. Along with this, there may be chat text that floats in a bubble above the avatar or appears in the chat window with the avatar's name; the avatar's name itself may float above the avatar; an avatar may have a greeting card or other objects to pass as part of its appearance. The avatar body is diverse: from seemingly photorealistic versions of a person, to furry monkeys, colorful dragons, silver skinned werewolves, M&M candies, milk bottles, and so on. The default body is human and chosen from a set of racial, gender, and cultural types. After doing so, it is easy to change the weight, skin tone, and so on. Attributes are sliding ranges. Height ranges from about 7'6" to 4'1". Sliders select eye color, amount of fat on the hips, size of ears, and so on. You easily customize your humanoid avatar into another gendered avatar. As the *Second Life* website says, "It's you—only in 3D. You can create an avatar that resembles your real life or create an alternate identity. The only limit is your imagination. Who do you want to be?"[7] The binary is crucial: real life or an alternative opposes logics of mirroring to that of fiction. In either case, an identity measured in relation to an absent subject. "The only limit is your imagination." Despite this sentence, the limit on the imagination is the prescription that you must be either yourself or an alternative. Be who you want to be, be an other. The only requirement is that *you must be*, and this is prescribed: make and individuate yourself as a sign.

The stories and ads on the *Second Life* website show images of avatars that closely resemble their owners or users. The site also features ads and stories of radical divergence, where users cross gender or choose non-human avatars. There is no requirement to be human, but it takes a bit of work to make the change. Re-making requires editing the default shape. All avatars, of any gender or species or without species or without gender, are built from this default shape. To become different requires distorting, stretching, and molding the default shape. The shape can be wrapped with a "mask" or graphic representing the head. You can add a skin texture to give the shape a tone, hair, tattoos, and so on. Along with this are clothes and other added body parts.

To become other than human requires more than just molding, stretching, and distorting.[8] The default shape is hollow and your labor works through this hollow. Pull. Distort. Delete. Non-human avatars add parts to the default shape and make other parts invisible. Start. Scan. Filter. Manipulate. Map. Of course, there is no need to scan, no need to start from a putative trace of the real. You can just buy a non-human shape for ten dollars or so, or find a free one. There is an economy and gift exchange of body parts, skin tones, prostheses, clickable poseballs with positions and animations, and so on.

These partial objects add up to the avatar "itself." The aggregate is not simply the appearance of the avatar but the assertion that "I appear," whatever the appearance—human, furry, dragon.... The apparatuses position and fix the avatar sign as a scene of communication. Through these apparatuses, you are at home in the avatar. Take this literally: at home in *Second Life* in the sense developed by Boellstorf's *Coming of Age in Second Life*.[9] There is perception on the screen, supplemented and verified by chat, by other data. The avatar and the subject, or you and the prim, are within each other. You consume each other.

Avatars are intentional, and involve practices and cultures supporting occluded inhabitations. The sign is an action. It is common knowledge that the term "avatar" comes from a Hindu word for human embodiment of a god, but it is often forgotten that such avatars were embodied for a purpose, as a mapping of an absent body and intention of otherness from outside the world. The avatar is an announcement from elsewhere. The performativity of the *Second Life* sign is stiff with this absence. For this reason, the absent puppeteer is the subject-that-persists within appearances, and the objects of the world are empty shells. The avatar communicates the absent subject-that-persists through the regular and stable readability of assemblages of signs. The avatar and dance must be, must mean.

The avatar body is balanced and contained. Its surfaces are numbered and coded. Even when the body is a mess, it is proportioned and cleaned. Even the most distorted avatar does not bleed. There is no fluid, no mucus, no yolk, no vomit, no slobber, no foam. No avatar touches. No avatar touches another avatar. No avatar touches me. I touch the screen and rub the pixels there, but I never touch the avatar. No avatar touches itself. Of course, there are endless lines of circle jerking avatars masturbating online. The script runs and avatar hands are busy, but no palm touches skin, no surface is palpitated, no clit is stroked. You may tug away on your side of the screen while the code executes on the other side.

There is hygiene and purity. Think of this as partitioning. On this side of the screen there is my breath, my touch, my drool across the keyboard. On that side of the screen, the avatar can flow, break, and ooze, all in perfect animated color. I can give nothing to the avatar. Not a caress, not a strike, not a word, not a shit. I couldn't give a shit for an avatar. I hold everything in or spew it without return. *Second Life* lives on.

As René Thom put it, "all semantics necessarily depends on a study of space—geometric or topological."[10] You must recognize that even the most distorted and odd-appearing avatar recreates and maintains postural schemata of control and bodily comportment. The components are the same: to be an avatar requires an internal skeleton and a mesh of surface textures. The control and movement of the avatar is the same as well. Being in *Second Life* is to be a skeleton with coordinate nodes and sculptable surfaces. To be a stain, a dragon, or a toaster is the same as to be an emo buxom porn fantasy woman, the only difference being the wrapping, the textured surface. Cultures are mapped and re-mapped but the way of inhabiting remains the same. As the Vedanta philosopher Madhvacharya held, all avatars are alike in power and in all qualities.

In fact, all things in *Second Life* are mappings in this way. Objects in the virtual world are built of generic stuff or matter, on the one hand, and mappings that are applied from outside, on the other. On the first hand, generic objects are formed of spatial coordinates and given volume and visibility. Such objects are the basic "stuff" of the world. On the second hand, objects are made into prims or "sculpted primitives" using UV texture maps. Such maps describe object's three-dimensional x,y,z coordinates in two-dimensional terms, or u,v. The result is a scalable description of vortex positions, light and shade, and so on. For example, a 2D texture map of a tree applied to a generic 3D object in *Second Life* is sculpted to become a prim with the dimensions and contours of a tree. The remapping of geometrically formed matter gives the feel of an organic world. The organic "feel" or stuff is the sacrament of the virtual world. To be more precise, prims, as amalgamations of re-mapped coordinates and data clouds, are the perfection of reality as virtual, of reality as container and gridding of matter. *Second Life* is for and against first life. It is the extraction, presentation, and representational triumph of first life. Not a specific virtual environment but an operation on life itself. Is this not what we always already are? Material frames with cultural imprints or injections, embodiment as remapping of culture?

Passion of the avatar, avatar of passion

The dance is lost, the dancer is elsewhere. The point is about intimacy. Intimacy is deep, close friendship. No, it is beyond friendship and into the erotic. I am writing about an intimacy with *Second Life*, and for that matter with the work of Alan Sondheim in *Second Life*. I am writing about what I see as a deep friendship but also a corporeal relation, where our avatar bodies do it. Do what? Do everything in *Second Life*. There, I said it, or rather wrote it. I wrote it as a saying or announcing. I intimated our relation. What relation? Between two avatars. To intimate is to impress and announce, but also to make a legal communication. Intimacy is sealed and bound. Intimations are utterances and phrases that bind the deep erotics of friendship. "I do" is an utterance that expresses intimacy as well as legalizes it. The end of intimacy is legally concluded through divorce or other proceedings. Language plays with contracts that bind the body and series of bodies.

But then the questions start. How do we start? Can I only be intimate with someone I am already intimate with? Does the seal of intimacy only confirm the relation already there? In a crowd, my eyes meet with someone for a moment and there is a shock, a tug in the spine, in the scrotum, a stirring in the lizard parts of the mind. Is this not intimacy? For that matter, how can I be intimate with an avatar? My text goes nowhere and comes nowhere. Or rather, coming and going without end or beginning. Nothing is figural among these disarticulations and dismemberments, these parts and remainders of distended surface. *Second Life* remains an enigma. Alan Dojoji and Sandy Taifun may dance, may spiral, may flutter as prayer flags in the haze. The figural meant the metaphor of metaphor, the promise of conceptuality grasped in the play of the figural, and finally the reflexivity on culture that this conceptuality proffers. The humming you hear, the visual noise you stare at, the narrative I have been telling you, the languages we read: nothing more or less than the perverse desire for *Second Life* to be a scene of power, to be a scene of body, to be a scene of my body, of my discourse, my control, my culture.

I can give nothing to the avatar. As I already wrote, I couldn't give a shit for an avatar. There is much to write about avatars—but the blockage, the barrier between myself and the virtual is absolute and solid. Which is to say, the barrier is on me, on the skin, on the trachea as I speak, on my intestine as I digest and excrete. I live the barrier, I am the barrier.

Second Life is not a world of matter and things but of arrivals, cuts, and interruptions. I can teleport between locations. I can switch my viewpoints. It is a world of many absences sewn together by my immediacy to announcements of the avatar as mobile sign array. To enter *Second Life* is to arrive at speed—sometimes falling in space—and to arrive as bodily cohesion. You suddenly drop where you last logged out. Sometimes the setting is different, things changed, certainly the other avatars are gone; those others you were intimate with are not logged on. Often there is a lag in login, as avatar bodies are slowly drawn, downloaded, clothed. Parts are filled in, turning from outlines to meshes of information, body and organs thrown together until they take on a semblance of resemblance. You cohere or inhabit, arrive in your body, almost as if cooking or maturing. Bodies enter and depart the world. Only a small number of the over 21 million *Second Life* users are logged in at any one time. The highest average for concurrent users logged in was about 54,000, and it is typically far less. All avatars are positioned in the world. In between logins, bodies hang invisible in the space, or at least the world retains their last position. To enter *Second Life* is also to enter at a time of day and location. It may be midnight or high noon, and while these conditions can be changed, entry into the world throws them suddenly at you. Such conditions are necessary and ineluctable to being in *Second Life*.

What experience is this like? By this, I ask our first life, of all our lives and fields of being: what is the experience of *Second Life*? To arrive and deepen inhabitation, to enter a shell and slowly thicken into a body. When do we enter and leave our world? Day to day, walking around, crossing thresholds. To be sure, there are breaks, there are transformations, but are they ever so clean as this, so abrupt? Are they ever such a formalized entry into existence? Are we ever such a skeleton of stuff slowly colored and cultured? Are we ever not in the world? Perhaps in sleep, delirium, or ecstasy there is a disturbance of our situation; perhaps in the deepest intimacy.

To experience extreme pleasure or the depths of pain may shift us out of ourselves, but perhaps every moment is inhabitation sealed by intimacy. Perhaps *Second Life* is this: nothing other than intimacy. To wake up in the morning is to wake up into a time and place, a dimension of light, a location on the earth. Existence as a drift or breeze, as a sensation and atmosphere, not as coordinates and skeletons, not as operational frameworks. Avatars of friendship, of love, of passion, of words to others.

NOTES

Introduction

1 There are many fine books of this sort. Start with Funkhouser, *New Directions*, and Hayles, *Electronic Literature*.

2 At the conclusion of his book on the German *Trauerspiel*, Walter Benjamin writes of an exaggerated or overextended transcendence that in fact deepens its effects in "manifest subjectivity" (234), which in turn is exactly the guarantee of the miracle. For my purposes, bodily delirium and ecstasy is our nodal relation in the network, a relation so subjective it can only be expressed in writings such as this book.

3 The network is "otherwise than being," in the sense Levinas develops in that necessary book.

4 Spolsky, 2002 unpaginated.

5 Sondheim, "Gender and You," 196.

6 I read and re-read Sondheim's "Internet Text" which begins, "The following iterative construct describes the internet. Traditional philosophical and cultural categories are highly problematic, given this new communications life-form. I develop instead an analysis based on EMISSION, SPEW, ADDRESS, RECOGNITION, and PROTOCOL. The essay, always preliminary, proceeds through a series of numbered paragraphs, which then rewrite." I attempt to write my response to this work, the most important I know about the Internet, in my introduction to Sondheim's *Writing Under*.

7 From the Introduction to Lingis' *Abuses*: "Finding the right words takes time, and the one to whom they are addressed is no longer the one you thought he or she was when you wrote. One sends one's letters to an address he or she has left."

Chapter I

1   The work of N. Katherine Hayles is typical here, using "inscription practices" as a general rubric for how technologies embody human subject and how embodied subjects work with technologies (see Hayles, *Electronic Literature*). The model is flexible and somewhat dialectical, though never entirely clear in the actual definition of inscription, an un-clarity that enables the flexibility.

2   For more, see my "Purple Dotted Underlines."

3   The event of my writing is where the real stops not being written, in the Lacanian sense.

4   Of course, all this implies Paul Virilio's "logistics of perception" in his works from 1980 onward, especially *War and Cinema*, *Lost Dimension*, and *The Vision Machine*.

5   See Stelarc, "The Cadaver, The Comotose, and the Chimera," unpaginated.

6   See Freud, *Beyond the Pleasure Principle*, 19 and 20.

7   Plato, *Parmenides*, unpaginated.

Chapter II

1   Chapman, *Meatphysics*, 1.

2   Chapman, *Meatphysics*, back cover.

3   Hayles, *Electronic Literature*, 43–86.

4   Sondheim, "l.txt," unpaginated.

5   Borenstein and Freed, "Mime," unpaginated.

6   W3, "Client Handling of MIME Headers," unpaginated.

7   Sondheim, "My Reply of Poetics," unpaginated.

8   Beckett, *Stories and Texts for Nothing*, 114.

9   Von Foerster, *Understanding Understanding*, 185.

Chapter III

1   Baran, *On Distributed Communications Networks*.

2   Baran, *On Distributed Communications*.

3   Galloway, *Protocol*.

4   Turing, *On Computable Numbers*.

5   See Figure 1, http://electronicbookreview.com/sites/default/files/ essays/ping1largeweb.jpg

6   See Figure 2, http://electronicbookreview.com/sites/default/files/ essays/ping2largeweb.jpg

7   For the use of crowds here and elsewhere, I imply Deleuze and Guattari's *A Thousand Plateaus* and, behind that, Canetti's *Crowds and Power*.

8   Winograd and Flores, *Understanding Computer and Cognition*.

9   Cerf and Kahn, "A Protocol for Packet Network Communication."

10 Spolsky, "Leaky Abstraction."

11 Braden, "Requirements for Internet Hosts."

12 Ashby, *An Introduction to Cybernetics*, 1.

13 Ashby, *An Introduction to Cybernetics*, 3.

14 Janicki, "Ping Melody."

15 V2 Archive, "Ping Body."

16 Lingis, "Society of Dismembered Body Parts," 297.

17 Bellmer, *Little Anatomy*, 26.

18 Bellmer, *Little Anatomy*, 36–37.

19 Thayer, "The Law Which Underlies Protective Coloration," 127.

20 Portman, *Animal Forms and Patterns*, Chapter VI.

21 Bellmer, *Little Anatomy*, 5–6.

22 Chasseguet-Smirgel, *Creativity and Perversion*, 36.

Chapter IV

1   Franz Kafka's "The Imperial Message" remains the perfect allegory of communications network: at the end, the message never reaches you but you dream it nonetheless.

2   Maturana, 1980.

3   Winograd and Flores, *Understanding Computer and Cognition*.

4   Luhmann, 1996.

5   See Tarski, *Logic, Semantics, Metamathematics*.

6   Thompson, "Reflections on Trusting Trust."

7   Mishima, *Sun and Steel*, 29.

8   Leder, *The Absent Body*.

9   Kristeva, *Reader*, 26.

10   Duncan, *Opening*, 7.

11   Ritchie and Thompson, "The Unix File-Sharing System," 2–3.

12   Lessig, *Remix*, 28.

13   Bal, *Narratology*, 5.

14   Benveniste, *Problems*, 223.

15   Blanchot, *Infinite Conversation*, 424.

Chapter V

1   Freud, *The Question of Lay Analysis*, 16.

2   Adorno, *Aesthetic Theory*, 261.

3   Adorno, *Negative Dialectics*, 17.

4   Wilden, *System and Structure*, 155–190.

5   Sartre, *Notebook for an Ethics*, 37.

6   Merleau-Ponty, *The Visible and the Invisible*, 130–155.

7   Gibson, *Idoru*, 178.

8   Sartre, *The Words*, 193.

9   Adorno, *Aesthetic Theory*, 261.

10   Lacan, *Écrits*, 247.

11   Here and elsewhere, when I refer to the literary community, I am implying the arguments of Blanchot, *The Unavowable Community* and Lingis, *The Community of Those Who Have Nothing in Common*.

12   Electronic Literature as a Model of Creativity in Practice.

13   Barthes, *Lover's Discourse*, 1.

14   "What is Electronic Literature?"

15   von Foerster, *Understanding Understanding*, 299.

16   I suggest looking at the academic institutions that are tied to the emergence of the field of electronic literature: Brown University, SUNY Buffalo, the University of Paris 8, and so on. The list is long and growing.

17   Bateson, *Steps*, 299.

18   Hayles, *Electronic Literature*, 4–5.

19   Duncan, "Rites of Participation," 327.

Chapter VI

1 Nietzsche, *Will to Power*, 367.
2 McCarty, "The Irony of Spam," unpaginated.
3 Stivale, "Spam, Heteroglossia, and Harassment."
4 Ryle, *The Concept of Mind*, 188.
5 Shirky, *Here Comes Everybody*, 88.
6 In Mike Musgrove, "Email Reply," unpaginated.
7 Sweet, *Inventing the Victorians*, 38–44.
8 Spamhaus Project, "Definition of Spam," unpaginated.
9 Reich, *Character Analysis*, 169–193.
10 Crocker, "The Standard for the Format of ARPA Internet Text Messages," unpaginated. This replaces the earlier RFC 733.
11 Agamben, *End of the Poem*, 110.
12 Duncan, *Ground Work: Before the War/In the Dark*, 5.
13 Morville, *Ambient Findability*, 4.
14 Blanchot, *Infinite Conversation*, 350.
15 Galloway and Thacker, *The Exploit*, 146.
16 Derrida, *The Postcard*, back cover.
17 Shannon, "Mathematical Theory," 2.
18 Thomas, "On the Problem of Signature Authentication for Network Mail," 1.
19 Thomas, "On the Problem of Signature Authentication for Network Mail," 1.
20 Sartre, *Being and Nothingness*, 615.
21 Sondheim, "net2.txt," unpaginated.
22 Nelson, "Proposal," 446.
23 Berry, *Greetings in the Name of Jesus*.

Chapter VII

1 Freud, *The Standard Edition*, 300. This is from a late set of notes gathered as "Findings, Ideas, Problems."
2 Freud, *The Question of Lay Analysis*, 16.
3 Crocker, "Host Software."

4   "IBM Clock Corner."

5   Flynt, "The Counting Stands."

6   Melnick, *Pcoet*, and Andrews, *I Don't Have any Paper*.

7   The link is dead, but the project was written up on *Boing Boing* by Doctorow, "CAPCHAs as Random Poetry."

8   Bartoll, "Are You Human?"

9   Stern, "CAPTCHA Paintings."

10 Gus23, "Dream CAPTCHA."

11 Quoted in Harper, *Philosophy of Time*, 25.

12 Dalkilic, et al. "Using Compression to Identify Classes of Inauthentic Text."

13 See the comic strip Dilbert of March 30, 2008, where Dilbert's boss states, "Well, it is what it is" and "You don't know what you don't know," and Dilbert congratulates him as "the first human to fail the Turing Test" (Adams, 2008). The combination of everyday dialogue and trite business-speak humorously stages the constant testing involved with being human.

14 von Ahn, "CAPTCHA, the ESP Game, and Other Stuff."

15 "What is ReCAPTCHA?"

16 Doctorow, "Solving and Creating CAPTCHAS with free porn." This practice may be apocryphal, but it demonstrates the connection between solving CAPTCHAs and desiring production.

17 "What is ReCAPTCHA?"

18 von Ahn, "Games with a Purpose."

19 "Amazon Mechanical Turk."

20 Mieszowski, "I Make $1.45 a Week and Love It."

21 Valla, "Sol Lewitt + Mechanical Turk."

22 Koblin, "The Sheep Market" and Koblin and Kawashima, "Ten Thousand Cents."

23 Lacan, *Écrits*, 247.

24 Turing, "Computing Machinery," 439.

25 Turing, "Computing Machinery," 439.

26 "Intel Processor."

27 May, "The Inaccessibility of CAPTCHA."

28 Naor, "Verification of a Human in the Loop," 2–3.

29 Lyotard, *The Inhuman.*

30 Bick, "The Experience of the Skin."

Chapter VIII

1   *Key Information Security Terms,* 142.

2   Shirley, "Internet Security Glossary," 126.

3   Friedman, *Elements of Cryptanalysis,* 138.

4   Foucault, *Archeology of Knowledge,* 88.

5   *American Standard Code,* Unpaginated Foreword.

6   *Archeology of Knowledge,* 88; *American Standard Code,* 6.

7   Sartre, *Critique of Dialectical Reason,* and Rosset, "Reality and the Untheorizable," 83.

8   Bachelard, *Water and Dreams,* 104–113.

9   Lacan, *Écrits,* 433.

10 "Acclaim for Unicode."

11 "What is Unicode?"

12 *The Unicode Standard,* 1.

13 *The Unicode Standard,* 1.

14 Lennon, *In Babel's Shadow,* 170.

15 Becker, *Unicode* 3.

16 *The Unicode Standard,* 11.

17 Korpela, *Unicode Explained,* 20.

18 "The Unicode Standard: A Technical Introduction," unpaginated.

19 Saussure, *Course,* 118.

20 Freud, *Interpretation of Dreams,* 261.

21 I came across Nicole Shukin's *Animal Capital: Rendering Life in Biopolitical Times* only after writing this text. Her work is an astute unpacking of the semantics of "rendering."

22 Giedion, *Mechanization Takes Command,* 240.

23 Giedion, *Mechanization Takes Command,* 246.

24 *The Unicode Standard,* 5.

25 McLuhan, *The Gutenberg Galaxy.*

26 Magnhildøen, "Plaintext Performance," unpaginated.

27 Magnhildøen, "Plaintext Performance," unpaginated.

28 Hayles, *Writing Machines*, 25–26.

29 Magnhildøen, "Plaintext Performance," unpaginated.

30 Friedman and Elizabeth, *The Shakespearian Ciphers Examined*.

31 Mac Low, *Stanzas for Iris Lezak*.

32 *American Heritage Word Frequency Book*, 601.

33 Selfridge and Neisser, "Pattern Recognition."

34 Friedman, *The Index of Coincidence*.

35 Shannon, "Mathematical Theory of Communication," 7.

36 Shannon, "Mathematical Theory of Communication," 7.

37 Shannon, "Mathematical Theory of Communication," 7.

38 Shannon, "Mathematical Theory of Communication," 7.

39 *American Heritage Word Frequency Book*, xiii.

40 Baldwin, "Process Window."

41 Bennett, "Logical Depth," 230.

42 Morton, "The Beginner's Guide to Pretty Good Privacy," unpaginated.

43 Shannon, "Communication Theory of Secrecy Systems," 662.

44 Kerkhoff summarized by Andress, *Basics of Information Security*, 68. Friedman's *Elements of Cryptanalysis*, the manual used for decades by U.S. Army cryptographers, formulates a similar principle on page 137, elevating computational complexity over secrecy as such: "The best that can be expected is that the system should be complicated enough to resist analysis for such a length of time that when solution is finally achieved, the information obtained is of no special value."

45 Kristeva, *Revolution in Poetic Language*, 19–90.

Chapter IX

1   "Second Life: Homepage," unpaginated.

2   Alan Sondheim and I have performed dozens of times over the last decade, most recently at the Centre Pompidou, but always in *Second Life*, always elsewhere.

3   "Second Life Community Standards," unpaginated.

4 Nakamura, *Cybertypes*, 523.

5 Chasseguet-Smirgel, *Creativity and Perversion*, 74–91.

6 Some information on Sondheim's *Second Life* work is available on his website, http://alansondheim.org.

7 "Second Life: What is an Avatar?" unpaginated.

8 Micha Cárdenas' project "Becoming Dragon" is a radical take on becoming other in *Second Life*. See http://www.ctheory.net/articles.aspx?id=639.

9 See Boellstorf, *Coming of Age in Second Life*.

10 Thom, *Mathematical Models*, 275.

BIBLIOGRAPHY

Abbate, Janet. Inventing the Internet. Cambridge: The MIT Press, 2000.

"Acclaim for Unicode." Unicode.org. http://www.unicode.org/
announcements/quotations.html. Accessed April 6, 2014.

Adams, Scott. Dilbert. Andrews McMeel Publishing, March 30, 2008.

Adorno, Theodor. Negative Dialectics. Translated by E. B. Ashton. New
York, NY: Routledge, 1973.

———. Aesthetic Theory. Translated by Robert Hullot-Kentor. New York,
NY: Continuum, 1997.

Agamben, Giorgio. The End of the Poem. Translated by Daniel Heller-
Roazen. Stanford: Stanford University Press, 1999.

Amazon Mechanical Turk. https://www.mturk.com/mturk/welcome.
Accessed September 1, 2013.

American Standard Code for Information Interchange. American
Standards Association. June 17, 1963. http://worldpowersystems.
com/J/codes/X3.4-1963. Accessed April 6, 2014.

Andress, Jason. The Basics of Information Security. Waltham, MA:
Syngress, 2011.

Andrews, Bruce. I Don't Have Any Paper So Shut Up (or, Social
Romanticism). San Francisco, CA: Sun and Moon, 1992.

Ashby, Hal. An Introduction to Cybernetics. New York, NY: John Wiley,
1956.

Bachelard, Gaston. Water and Dreams: An Essay on the Imagination
of Matter. Translated by Edith R. Farrell. Dallas, NY: The Pegasus
Foundation, 1983.

Baldwin, Sandy. "Process Window. Code Work, Code Aesthetics, Code
Poetics." In The Cybertext Yearbook 2002–3. Edited by Markku
Eskelinen, Raine Koskimaa, Loss Pequeño Glazier and John Cayley.
Jyväskylä: JY Publications, 2003, 107–119.

———. "Purple Dotted Underlines: Microsoft Word and The End of
Writing." Afterimage. Volume 30. Number 1. 2002. 6–7.

Bal, Meike. Narratology: Introduction to the Theory of Narrative.
Toronto: University of Toronto Press, 1985.

Baran, Paul. On Distributed Communications Networks. Santa Monica,
CA: The Rand Corporation, 1962.

———. *On Distributed Communications: I. Introduction to Distributions Communications Networks*. Santa Monica: The Rand Corporation, 1964.

Barthes, Roland. *A Lovers' Discourse*. Translated by Richard Howard. New York, NY: Hill and Wang, 1978.

Bartoll, Aram. "Are you Human? Object and Urban Intervention." *datenform.de*. http://www.datenform.de/areyouhumaneng.html. Accessed September 1, 2013.

Bateson, Gregory. *Steps Towards an Ecology of Mind*. Chicago: University of Chicago Press, 1972.

Becker, Joseph. *Unicode*. Palo Alto, CA: Xerox Corporation, 1988.

Beckett, Samuel. *Stories and Texts for Nothing*. New York, NY: Grove Press, 1967.

Bellmer, Hans. *Little Anatomy of the Physical Unconscious, or The Anatomy of the Image*. Translated by Jon Graham. Waterbury Center, CT: Dominion, 2004.

Benjamin, Walter. *The Origin of German Tragic Drama*. Translated by John Osborne. London: Verso Books, 1998.

Bennett, Charles. "Logical Depth and Physical Complexity." In *The Universal Turing Machine: A Half-Century Survey*. Edited by Rolf Herken. Oxford: Oxford University Press, 1980.

Benveniste, Émile. *Problems in General Linguistics*. Miami: University of Miami Press, 1971.

Berry, Michael. *Greetings in the Name of Jesus! The Scambaiter Letters*. New York, NY: Harbour Books, 2006.

Bick, Esther. "The Experience of the Skin in Early Object Relations." In *Parent-Infant Psychodynamics: Wild Things, Mirrors, and Ghosts*. Edited by Joan Raphael-Leff. London: Whurr Publications, 2003.

Blanchot, Maurice. *The Unavowable Community*. Translated by Pierre Joris. Barrytown: Station Hill Press, 1998.

Boellstorff, Tom. *The Infinite Conversation*. Translated by Susan Hanson. Minneapolis: University of Minnesota Press, 1993.

———. *Coming of Age in Second Life: An Anthropologist Explores the Virtually Human*. Princeton: Princeton University Press, 2010.

Borenstein, N. and N. Freed. "MIME (Multipurpose Internet Mail Extensions): Mechanisms for Specifying and Describing the Format of Internet Message Bodies)." *IETF*. https://www.ietf.org/rfc/rfc1341.txt. Accessed April 1, 2014.

Braden, R. "Requirements for Internets Hosts." RFC #1. *IETF*. http://tools.ietf.org/html/rfc1122. Accessed April 1, 2014.

Canetti, Elias. *Crowds and Power*. Translated by Carol Stewart. New York, NY: Farrar, Straus and Giroux, 1984.

Cardenas, Micha. "Becoming Dragon." *CTheory.net*. http://www.ctheory. net/articles.aspx?id=639. Accessed April 29, 2010.

Carroll John Bissell, Peter Davies and Barry Richman, eds. *The American Heritage Word Frequency Book*. New York, NY: Houghton Mifflin, 1971.

Cerf Vint and Robert Kahn. "A Protocol for Network Intercommunication." *IEEE Transactions on Communications*. Volume 22, Number 5. May 1974. 637–648.

Chapman, Jake. *Meatphysics*. London: Creation Books, 2003.

Chasseguet-Smirgel, Janine. *Creativity and Perversion*. New York, NY: Norton, 1984.

Crocker, David H. "The Standard for ARPA Internet Text Messages." RFC #822. *IETF*. http://www.ietf.org/rfc/rfc0822.txt. Accessed April 6, 2014.

Crocker, Steve. "RFC #1: Host Software." *IETF*. 1969. http://www.ietf. org/rfc/rfc0001.txt?number=1. Accessed September 1, 2013.

Dalkilic Mehment, Wyatt T. Clark, James C. Costello and Predrag Radivojac. "Using Compression to Identify Classes of Inauthentic Text." *Proceedings of the Sixth SIAM International Conference on Data Mining*. Edited by Joydeep Ghosh. SIAM. 2006. https://mail.uhv. edu/exchweb/bin/redir.asp?URL=http://www.siam.org/meetings/sdm06/ proceedings/070dalkilicm.pdf. Accessed September 1, 2013.

Deleuze, Gilles and Félix Guattari. *A Thousand Plateaus*. Translated by Brian Massumi. Minneapolis: The University of Minnesota Press, 1987.

Derrida, Jacques. *The Post Card*. Translated by Alan Bass. Chicago: University of Chicago Press, 1987.

Doctorow, Cory. "CAPCHAs as Random Poetry." *Boing Boing*. http:// www.boingboing.net/2003/11/09/captchas_as_random_p.html. Accessed September 1, 2013.

———. "Solving and Creating CAPTCHAs with Free Porn." *Boing Boing*. http://www.boingboing.net/2004/01/27/solving_and_creating.html. Accessed September 1, 2013.

Duncan, Robert. *The Opening of the Field*. New York, NY: New Directions, 1960.

———. "Rites of Participation." In *Symposium of the Whole*. Edited by Jerome Rothenberg and Diane Rothenberg. Berkeley: University of California Press, 1983.

———. *Ground Work: Before the War/In the Dark*. New York, NY: New Directions, 2006.

"What is electronic literature?" *Electronic Literature Organization*. http:// eliterature.org/what-is-e-lit/. Accessed September 1, 2013.

Electronic Literature as a Model of Creativity in Practice. *ELMCIP.net.* Accessed April 6, 2014.

Flynt, Henry. "The Counting Stands: Plurality, Thinghood, Contradiction: An Instruction Manual for Conceptual Art." *Henryflynt.org.* http://www.henryflynt.org/meta_tech/thinghd.html. Accessed September 1, 2013.

Foucault, Michel. *The Archeology of Knowledge.* Translated by A. M. Sheridan Smith. New York, NY: Vintage, 1982.

Freud, Sigmund. *Beyond the Pleasure Principle.* Translated by James Strachey. New York, NY: Norton, 1961a.

———. *The Interpretation of Dreams.* Translated by James Strachey. New York, NY: Norton, 1961b.

———. *Standard Edition of the Complete Psychological Works of Sigmund Freud.* Translated by James Strachey. Volume 23. London: Hogarth Press, 1964.

———. *The Question of Lay Analysis.* Translated by James Strachey. New York, NY: Norton, 1969.

Friedman, William F. *Elements of Cryptanalysis.* Laguna Hills, CA: Aegean Park Press, 1976.

———. *The Index of Coincidence and its Application to Cryptography.* Laguna Hills, CA: Aegean Park Press, 1987.

Friedman, William F. and Elizabeth S. *The Shakespearian Ciphers Examined: An Analysis of Cryptographic Systems Used as Evidence That Some Author Other Than William Shakespeare Wrote the Plays Commonly Attributed to Him.* Cambridge: Cambridge University Press, 1957.

Funkhouser, C. T. *New Directions in Digital Poetry.* New York, NY: Continuum, 2012.

Galloway, Alexander. *Protocol: How Control Exists After Decentralization.* Cambridge, MA: The MIT Press, 2006.

Galloway, Alexander and Eugene Thacker. *The Exploit: A Theory of Networks.* Minneapolis: University of Minnesota Press, 2007.

Gibson, William. *Idoru.* New York, NY: Berkley Books, 1997.

Giedion, Sigfried. *Mechanization Takes Command: A Contribution to Anonymous History.* New York, NY: Norton, 1969.

Gus23. "Dream Captcha #1." *Gus23.* http://gus23.wordpress.com/2008/10/07/new-work-dream-captcha. Accessed September 1, 2013.

Harper, Albert. *The Philosophy of Time.* Ann Arbor, MI: E. Mellen Press, 1990.

Hayles, N. Katherine. *Writing Machines.* Cambridge, MA: MIT Press, 2002.

———. *Electronic Literature: New Horizons for the Literary*. Notre Dame: Notre Dame University Press, 2008.

"IBM Clock Corner Reference." *IBM Archives*. http://www-03.ibm.com/ibm/history/exhibits/cc/cc_room.html. Accessed September 1, 2013.

"Intel Pentium Processor Thermal Design Guidelines." Intel Corporation. 2009. http://www.intel.com/support/processors/pentium/sb/cs-011039.htm. Accessed September 1, 2013.

Janicki, Pawel. "Ping Melody." *Pawel Janicki*. http://paweljanicki.jp/pingmelody_main_en.html. Accessed April 1, 2014.

Kissel, Royed. *Glossary of Key Information Security Terms*. Washington: National Institute of Standards and Technology, 2013

Koblin, Aaron. "The Sheep Market." Aaron Koblin. 2010. http://www.aaronkoblin.com/work/thesheepmarket. Accessed April 1, 2014.

Koblin, Aaron and Takashi Kawashima. Ten Thousand Cents. 2010. http://www.tenthousandcents.com. Accessed April 1, 2014.

Korpela, Jukka. *Unicode Explained*. Sebatapol, CA: O'Reilly Media, 2008.

Kristeva, Julia. *Revolution in Poetic Language*. Translated by Margaret Waller. New York, NY: Columbia University Press, 1984.

Kristeva, Julia. *The Kristeva, Reader*. edited by Toril Moi. New York, NY: Columbia University Press, 1986.

Lacan, Jacques. *Écrits*. Translated by Bruce Fink. New York, NY: Norton, 2006.

Leder, Drew. *The Absent Body*. Chicago: University of Chicago Press, 1990.

Lennon, Brian. *In Babel's Shadow: Multilingual Literature, Monolingual States*. Minneapolis: University of Minnesota Press, 2010.

Lessig, Lawrence. Remix. *Remix: Making Art and Commerce Thrive in the Hybrid Economy*. New York, NY: Penguin Press, 2008.

Levinas, Emmanuel. *Otherwise than Being*. Translated by Alphonso Lingis. Pittsburgh, PA: Duquesne University Press, 1998.

Lingis, Alphonso. *The Community of Those Who Have Nothing in Common*. Bloomington: Indiana University Press, 1994a.

———. "The Society of Dismembered Body Parts." In *Deleuze and the Theater of Philosophy*. Edited by Constantin V. Boundas and Dorothea Olkowski. New York, NY: Routledge, 1994b.

———. *Abuses*. Berkeley: University of California Press, 1995.

Luhmann, Niklas. *Social Systems*. Translated by John Bednarz. Stanford: Stanford University Press, 1996.

Lyotard, Jean-François. *The Inhuman: Reflections on Time*. Stanford: Stanford University Press, 1992.

Mac Low, Jackson. *Stanzas for Iris Lezak*. New York, NY: Something Else Press, 1971.

Magnhildøen, Bjørn. "Plaintext Performance." *The Electronic Literature Collection Volume 2*. The Electronic Literature Organization. 2011. http://collection.eliterature.org/2/works/magnhildoen_ plaintextperformance.html. Accessed April 6, 2014.

Maturana Humberto and Varela Francisco. *Autopoiesis and Cognition*. Dordecht: D. Reidel Publishing, 1980.

May, Matt. "Inaccessibility of CAPTCHA." *World Wide Web Consortium*. http://www.w3.org/TR/turingtest. Accessed September 1, 2013.

McCarty, Willard. "The Irony of Spam." *The Humanist Discussion Group*. June 14, 2003. http://dhhumanist.org/Archives/Virginia/ v17/0083.html. Accessed April 6, 2014.

McLuhan, Marshall. *The Gutenberg Galaxy: The Making of Typographic Man*. Toronto: University of Toronto Press, 1962.

Melnick, David. *PCOET*. San Francisco, CA: G. A. W. K, 1975.

Merleau-Ponty, Maurice. *The Visible and the Invisible*. Translated by Alphonso Lingis. Chicago: Northestern University Press, 1968.

Mieszowski, Katherine. "I Make $1.45 a Week and Love It." *Salon*. June 24, 2006. http://www.salon.com/2006/07/24/turks_3/. Accessed July 24, 2006. Unpaginated.

Mishima, Yukio. *Sun & Steel*. Translated by John Bester. New York, NY: Grove Press, 1970.

Morton, Bill. "The Beginner's Guide to Pretty Good Privacy." *Resist Security*. April 13, 1995. http://security.resist.ca/bg2pgp.shtml. Accessed April 6, 2014.

Morville, Peter. *Ambient Findability: What We Find Changes Who We Become*. Sebastapol, CA: O'Reilly, 2005.

Musgrove, Mike. "Email Reply: 'Leave me alone.' " *The Washington Post*. May 25, 2007. http://www.washingtonpost.com/wp-dyn/content/ article/2007/05/24/AR2007052402258_pf.html. Accessed April 6, 2014.

Nakamura, Lisa. *Cybertypes: Race, Ethnicity, and Identity on the Internet*. New York, NY: Routledge, 2002.

Naor, Moni. "Verification of a Human in the Loop or identification via a Turing Test." The Weizmann Institute. 1996. http://www.wisdom. weizmann.ac.il/~naor/PAPERS/human.pdf. Accessed September 1, 2013.

Nelson, Theodor H. "Proposal for a Universal Electronic Publishing System and Archive." In *The New Media Reader*. Edited by Noah Wardrip-Fruin and Nick Monfort. Cambridge: The MIT Press, 2003.

Nietzsche, Friedrich. *The Will to Power*. Translated Walter Kauffman. New York, NY: Vintage Books, 1968.

Plato, Parmenides. Translated by Benjamin Jowett. *The Internet Classics Archive*. http://classics.mit.edu/Plato/parmenides.html. Accessed March 23, 2014.

Portmann, Alfred. *Animal Forms and Patterns*. Translated Hella Czech. New York, NY: Schocken, 1967.

Reich, Wilhelm. *Character Analysis*. Translated by Vincent R. Carfagno. New York, NY: Farrar, Straus, Giroux, 1972.

Ritchie, D. M. and K. Thompson. "The Unix File Sharing System." *The Bell System Technical Journal*. Volume 57. Number 6. part 2. July–August 1978. http://www.cs.berkeley.edu/~brewer/cs262/unix.pdf, 365–375.

Rosset, Clement. "Reality and the Untheorizable." In *The Limits of Theory*. Edited by Thomas M. Kavanagh. Stanford: Stanford University Press, 1989.

Ryle, Gilbert. *The Concept of Mind*. Chicago: University of Chicago Press, 2000.

Sartre, Jean-Paul. *The Words: The Autobiography of Jean-Paul Sartre*. Translated by Bernard Frechtman. New York, NY: Vintage, 1983.

———. *Notebook for an Ethics*. Translated David Pellauer. Chicago: University of Chicago Press, 1992.

———. *Being and Nothingness*. Translated by Hazel E. Barnes. New York, NY: Washington Square Press, 1993.

———. *Critique of Dialectical Reason*. Translated by Alan Sheridan-Smith. New York, NY: Verso Books, 2004.

Saussure, Ferdínand de. *Course in General Linguistics*. Translated by Roy Harris. Peru, IL: Open Court, 1986.

"Second Life: Community Standards." *Secondlife.com*. Second Life website. Accessed April 6, 2014.

"Second Life: Home Page." *Secondlife.com*. Second Life website. Accessed April 6, 2014.

"Second Life: What is an Avatar?" *Secondlife.com*. Second Life website. Accessed April 6, 2014.

Selfridge, Oliver G. and Ulrich Neisser. "Pattern Recognition by Machine." *Scientific American*. Volume 23. 1960. 60–68.

Shannon, Claude. "A Mathematical Theory of Communication." *The Bell Systems Technical Journal*. Volume 27. July–October 1948. 623–656.

———. "Communication Theory of Secrecy Systems." Declassified government report. September 1, 1946. http://netlab.cs.ucla.edu/wiki/files/shannon1949.pdf. Accessed April 6, 2014.

Shirky, Clay. *Here Comes Everybody: The Power of Organizing Without Organizations*. New York, NY: Penguin Books, 2008.

Shirley, R. "Internet Security Glossary." RFC 2828. *Ietf.org*. http://www.ietf.org/rfc/rfc2828.txt. Accessed April 6, 2014.

Shukin, Nikole. *Animal Capital*. Minneapolis: University of Minnesota Press, 2009.

Sondheim, Alan. "Gender and You." *Transforming Cultures eJournal.*
Volume 2. Number 2. December 2007. Sydney: UTS Press, 2007.
194–200.

———. *Writing Under: Selections from the Internet Text.* Morgantown:
Computing Literature, 2012.

———. "l.txt." *The Internet Text.* http://collection.eliterature.org/1/works/
sondheim__internet_text/l.txt. Accessed April 1, 2014.

———. "My Reply of Poetics." *Alansondheim.org.* http://www.
alansondheim.org/prosepoetry.txt. Accessed April 1, 2014.

———. "net2.txt." *The Internet Text.* http://collection.eliterature.org/1/
works/sondheim__internet_text/l.txt. Accessed April 1, 2014.

The Spamhaus Project. "The Definition of Spam." *Spamhaus.org.* http://
www.spamhaus.org/consumer/definition/. Accessed April 6, 2014.

Spolsky, Joel. "The Law of Leaky Abstraction." *Joel on Software.* http://
www.joelonsoftware.com/articles/LeakyAbstractions.html. Accessed
November 11, 2002.

Spolsky, Joel. "The Law of Leaky Abstraction." *Joel on Software.* http://
www.joelonsoftware.com/articles/LeakyAbstractions.html. Accessed
March 31, 2014.

Stelarc. "The Cadaver, The Comotose, and the Chimera." *Stelarc.org.*
http://stelarc.org/documents/StelarcLecture2009.pdf. Accessed March
31, 2014.

Stern, Becky. "Captcha Paintings." *Sternlab.* http://sternlab.org/2008/07/
captcha-paintings. Accessed September 1, 2013.

Stivale, Charles. "Spam, Heteroglossia, and Harassment." In *Internet
Culture.* Edited by David Porter. New York, NY: Routledge, 1997.
Pages 133–144.

Sweet, Matthew. *Inventing the Victorians.* New York, NY: St Martin's
Press, 2001.

Tarski, Alfred. *Logic, Semantics, Metamathematics, papers from 1923
to 1938.* Edited by John Corcoran. Indianapolis: Hackett Publishing
Company, 1983.

Thayer, Abbott H. "The Law Which Underlies Protective Coloration."
In *The Auk.* Edited by J. A. Allen. Volume XIII. New York, NY:
L. S. Foster, 1896.

"Thermal Guide for the Boxed Intel Celeron Processor." *Intel.com.* http://
www.intel.com/cd/channel/reseller/asmo-na/eng/products/mobile/
processors/celeron/tech/381128.htm

Thom, René. *Mathematical Models of Morphogenesis.* Translated by
W. M. Brookes. Hemstead: Ellis Horwood, Ltd., 1983.

Thomas, Bob. "On the Problem of Signature Authentication for Network
Mail." RFC #644. *IETF.* http://www.ietf.org/rfc/rfc0644.txt. Accessed
April 6, 2014.

Thompson, Ken. "Reflection on Trusting Trust." *Communications of the ACM.* Volume 27. Number 8. August 1984. 261–263.

Turing, A. M. "On Computable Numbers: With an Application to the Entscheidungsproblem." *Proceedings of the London Mathematical Society.* Volume 43, Series 2, 1937. 230–265.

Turing, A. M. "Computing Machinery and Intelligence." *Mind.* Volume 49, 1950. 433–460.

The Unicode Standard 6.2. The Unicode Consortium. 2012. http://www.unicode.org/versions/Unicode6.2.0/UnicodeStandard-6.2.pdf. Accessed April 6, 2014.

"The Unicode Standard: A Technical Introduction." The Unicode Consortium. http://www.unicode.org/standard/principles.html. Accessed April 6, 2014.

Valla, Clement. "Sol Lewitt + Mechanical Turk." *Clement Valla.* 2008. http://www.clementvalla.com/index.php?/printed/sol-lewitt-mechanical-turk. Accessed September 1, 2013.

Virilio, Paul. *The Vision Machine.* Translated by Julie Rose. Bloomington: Indiana University Press, 1994.

———. *War and Cinema.* Translated by Patrick Camiller. London: Verso Books, 2009.

von Ahn, Luis. "CAPTCHA, the ESP Game, and Other Stuff." 2004. http://www.cs.cmu.edu/~biglou/cycles.ppt. Accessed September 1, 2013.

———. "Games with a Purpose." IEEE 44.6 (June 2006): 96–98. http://www.cs.cmu.edu/~biglou/ieee-gwap.pdf. Accessed September 1, 2013.

von Ahn Luis, Manuel Blum, Nicholas J. Hopper and John Langford. "CAPTCHA: Using Hard AI Problems for Security." *Eurocrypt.* 2003. http://www.cs.cmu.edu/~mblum/research/pdf/captcha.pdf. Accessed September 1, 2013.

Von Foerster, Heinz. *Understanding Understanding: Essays on Cybernetics and Cognition.* New York, NY: Springer-Verlag, 2003.

V2 Archive. "Ping Body." http://v2.nl/archive/works/ping-body. Accessed April 1, 2014.

"What is reCAPTCHA?" *reCAPTCHA.* http://recaptcha.net/learnmore.html. Accessed September 1, 2013.

"What is Unicode?" *Unicode.org.* http://www.unicode.org/standard/WhatIsUnicode.html. Accessed April 6, 2014.

Wilden, Anthony. *System and Structure: Essays in Communication and Exchange.* London: Tavistock, 1982.

Winograd, Terry and Francisco Flores. *Understanding Computer and Cognition.* Norwood, MA: Albex, 1986.

World Wide Web Consortium. "Client Handling of MIME Headers." *W3C.* http://www.w3.org/2001/tag/doc/mime-respect-20030505.html. Accessed April 1, 2014.

INDEX